Investigating the Ocean

An Interactive Guide to the Science of Oceanography

Second Edition

R. Mark Leckie

Richard Yuretich

University of Massachusetts, Amherst

The McGraw-Hill Companies, Inc.
Primis Custom Publishing

New York   St. Louis   San Francisco   Auckland   Bogotá
Caracas   Lisbon   London   Madrid   Mexico   Milan   Montreal
New Delhi   Paris   San Juan   Singapore   Sydney   Tokyo   Toronto

McGraw-Hill Higher Education

A Division of The McGraw-Hill Companies

Investigating the Ocean
An Interactive Guide to the Science of Oceanography

McGraw-Hill's Primis Custom Series consists of products that are produced from camera-ready copy. Peer review, class testing, and accuracy are primarily the responsibility of the author(s).

2 3 4 5 6 7 8 9 0 FPI FPI 0 9 8 7 6 5 4 3 2 1 0

ISBN 0-07-245845-3

Editor: Cynthia Biron
Cover Images: NOAA/NGDC
Cover Design: Joshua B. Staudinger
Printer/Binder: FirstPublish, Inc.

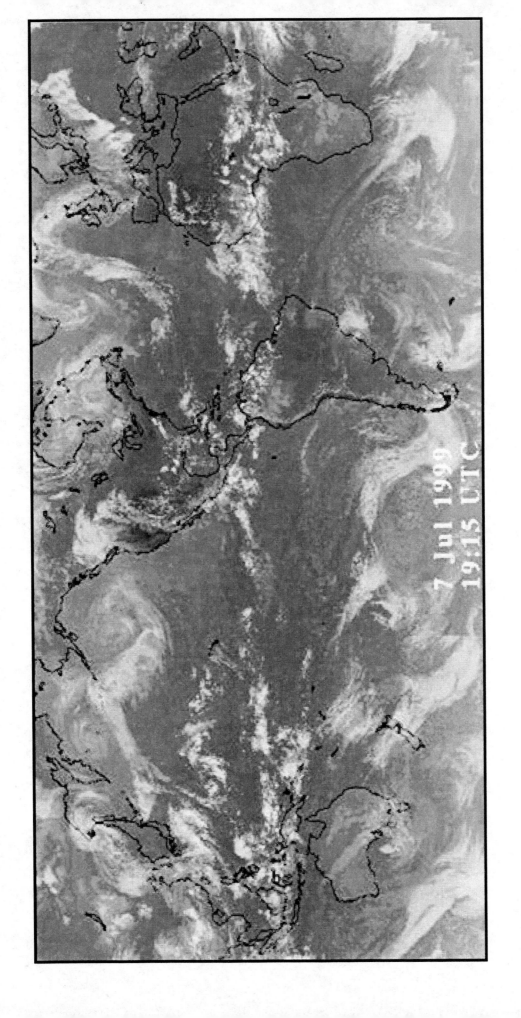

7 Jul 1999
19:15 UTC

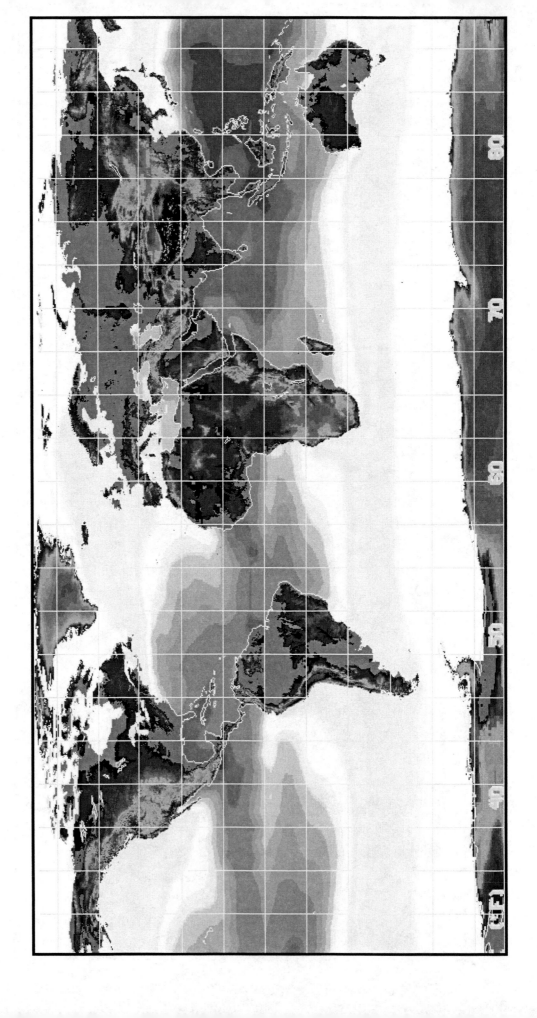

About this book

Educational research has shown that "experience is the best teacher," that is, our most memorable and long-lasting learning comes out of things we have done ourselves. The evidence that scientists believe this implicitly comes from the training of Ph.D. students, which is grounded fundamentally in guided discovery through research. Many colleges have also recognized the value of this approach for advanced undergraduates as witnessed by the senior projects or honors theses that are part of an increasing number of academic programs. However, introductory courses, especially those designed for the general student population, are only rarely influenced by these ideas. The traditional lecture and laboratory (or lecture only) is the way most students at colleges and universities fulfill their science requirements, and the experience is not always uplifting. This book is our attempt to help change the approach to teaching these courses.

Introductory Oceanography at the University of Massachusetts is a large course: approximately 600 students each semester divided into two lecture sections. There are no laboratories. Over the years we have developed several techniques to improve the learning environment in this class. The 75-minute class meeting is divided into segments consisting of short lectures interspersed with group investigations, problem-solving, interpreting scientific data and critical analysis of videos. The student-active portion of the class is incorporated into the worksheets in this book. These are designed to be interpretive and not "cookbook", to give students the time to contemplate the problem and deduce reasonable answers. In many cases the questions or discussion points are sufficiently flexible so that the instructor can use his or her favorite resources as a lead-in or follow-up to them. There are also several blank exercise pages so instructors can design their own investigations to meet the needs of the course. We use the exercises for short group investigations during the class, but they are equally suitable as homework assignments. Our approach has been deliberately "low-tech", access to the World Wide Web or sophisticated instrumentation is not required for any exercise. "Interactive" in the book title refers to the dialogue occurring among the students as well as between the class and the instructor. However, web sites and instructional technology can be very helpful in enriching the exercises.

We have also recognized a need for a concise overview of the major concepts in oceanography. Most modern textbooks are packed with information and contain beautiful pictures, but their encyclopedic approach often makes it hard to separate essential principles from supporting information. The second part of this book lays out the most significant aspects of oceanography, with each topic organized into in a two-page format. The page on the left is mostly text; the page on the right is primarily graphic. The two approaches complement and support each other so both textual and visual learners can benefit equally. Information to help with the exercises in the first part of the book can also be found here, and instructors can use the summaries and diagrams as the basis for their own planning.

This book is not intended to be a stand-alone text. There are many outstanding introductory oceanography textbooks on the market today, and over the years we have used a number of these books at the University of Massachusetts. Although we developed this book for a very large class, we believe it can be used equally well in smaller settings. The questions in the exercises can be discussed more thoroughly in smaller classes, many of the activities can be expanded, and the students can receive more extensive feedback on their responses. Whether the course is "lecture" only, or includes a separate laboratory section, this in-class book was designed to engage the students and encourage their active participation in classroom discussions. In any event, we hope that this book will stimulate more college instructors to use an investigative approach to teaching introductory oceanography.

We wish to acknowledge the contributions of Professors Laurie Brown and Julie Brigham-Grette of the University of Massachusetts, who also use these techniques in teaching this course and have contributed to the development of the in-class activities. The revision of this book has benefited from the comments of many individuals including Charles Bahr, Cindy Fisher, Michael McCormick, Desiree Polyak, Nicole Ramocki, and Lynda Vallejo. A special note of thanks is extended to Steve Nathan of the University of Massachusetts, Dr. Gretchen Andreasen of the University of California at Santa Cruz, and Professor Glenn Jones of Texas A&M University-Galveston, for their thorough reviews of the first edition and their insightful comments. The STEMTEC Project, a Collaborative for Excellence in Teacher Preparation of the National Science Foundation (DUE 9653966), has supported the production of these resources and the student-active approach to learning oceanography.

R. Mark Leckie
Richard Yuretich
University of Massachusetts
Amherst, MA 01003
July, 2000

To the student:

This book is a resource manual and guide that will help you learn about the oceans, explore some of the major phenomena that occur on our planet, and appreciate the way that scientific investigation of the Earth proceeds. The book is divided into two parts. The first part contains a series of short investigations that are designed to help you learn about a particular topic. Scientific knowledge is gained by exploration and discovery, that is, by individuals and teams working together to solve problems and pursue ideas. Accordingly, you really can't appreciate or understand the process by which we gather scientific information simply by listening to lectures and reading a textbook. You can engage in the discovery process by becoming involved in in-class discussions, experiments and video "field trips" as guided by the worksheets contained in this volume. Some of these worksheets will be used to assist with investigations during class time; others may be used as a basis for homework assignments. Still others will be used in conjunction with videos or films, so that you can maximize your understanding of the concepts that are presented in them. In this updated edition, we have assembled 37 different exercises dealing with many different aspects of oceanography. You will not do them all, but you may find that the worksheets not used in class may also help you in studying the topic.

The second part of the book is a concise overview of the major concepts of oceanography, which can serve as an additional resource to help you interpret the workings of the oceanic system. Several of the exercises refer to diagrams in this part of the book. Information that can help you complete your investigations can also be found here. You will also find this part of the book very helpful for study and review. Although most of the information is in your textbook, having a supplementary source that restates concepts with different words or an alternative approach can often be helpful in summarizing the major ideas of the topic.

To make effective use of this volume, **YOU MUST BRING THIS BOOK TO EVERY CLASS!** The worksheets are perforated near the binding, and you will frequently turn in completed worksheets at the end of a class meeting. We have deliberately kept the book small so you should have no problem carrying it in your backpack!

TABLE OF CONTENTS

Part 1: *Investigations*

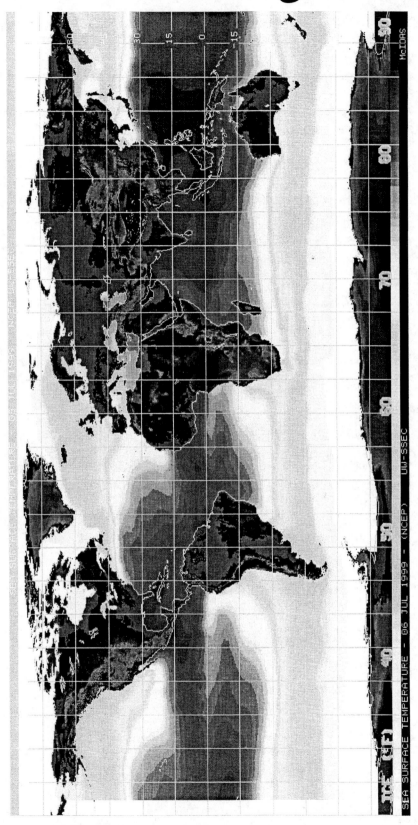

Additional Resources

The exercises that follow are designed to be flexible, so that they can be used in conjunction with a wide array of audiovisual, print and Internet resources. We have found that some videos and films can be used especially well with certain exercises.

#4: Continents and Ocean Basins. A video illustrating the basic principles of isostasy is available from the authors (currently being revised).

#12: Density. A video about measuring density is being produced by the authors.

#25: The Gulf Stream. The video *Gulfstream* (1982, National Film Board of Canada), is highly recommended.

#26: The Intertidal Zone. *The Intertidal Zone* (1985, National Film Board of Canada) is a good resource.

#27: Coastal Ecology. *Where the Bay Becomes the Sea* (1985, National Film Board of Canada) provides an excellent framework for discussion.

#30: Beaches. The film or video *Beach: a River of Sand* (1965, EBF) remains a very good summary.

#33: Coastal Erosion. *Portrait of a Coast* (1980, Circle Films) provides a complete overview of the problems involved.

#35: Overfishing: The video *The Last Hunters: the Cod War* (1990, Films for Humanities and Sciences) is a stimulating source for debate.

#36: Marine Pollution: *The Ocean Sink* (1990, Films for Humanities and Sciences) states some of the issues very well.

Clearly print your name:

Section:

Student number:

First 3 letters of last name

| #1 | Latitude and Longitude |

Latitude and Longitude are the principal means by which we can locate places on the globe. You will be given a subset of the following places. Put them in their correct locations to the nearest degree on the globes on the following page and on page 1 of this book. NOTE: Each place will be plotted twice on the following page, in either the Western or Eastern Hemisphere, and then either in the Northern or Southern Hemisphere.

	Place	Latitude	Longitude
	Europe		
1	Greenwich	51° 30' N	00° 00' W
2	Paris	48° 50'	02° 20' E
3	Madrid	40° 25' N	03° 45' W
4	Rome	41° 54' N	12° 29' E
5	Athens	37° 58' N	23° 46' E
6	Oslo	59° 55' N	10° 45' E
7	Reykjavik	64° 10' N	21° 57' W
8	Moscow	55° 45' N	37° 35' E
	Asia		
9	Ankara	39° 57' N	32° 54' E
10	Jerusalem	31° 47' N	35° 10' E
11	Cairo	30° 01' N	31° 14' E
12	Shanghai	31° 15' N	121° 26'E
13	Tokyo	35° 45'	139° 45'E
14	Baghdad	33° 20' N	44° 30' E
15	Bombay	18° 55' N	72° 50' E
16	Bangkok	13° 45' N	100° 35'E
	Australia & Oceania		
17	Sydney	33° 53' S	151° 10'E
18	Perth	31° 57' S	115° 52'E
19	Auckland	36° 52' S	174° 46'E
20	Fiji	17° 20' S	179° 00'E

	Place	Latitude	Longitude
	N. America		
21	Chicago	41° 53' N	87° 38' W
22	New York	40° 45' N	74° 00' W
23	Los Angeles	34° 04' N	118° 15'W
24	Calgary	51° 00' N	114° 10'W
25	Gander	48° 58' N	54° 35' W
26	Mexico City	19° 20' N	99° 10' W
27	Havana	23° 08' N	82° 22' W
28	Managua	12° 06' N	86° 20' W
	S. America		
29	Rio de Janeiro	23° 00' S	43° 12' W
30	Lima	12° 00' S	77° 00' W
32	Santiago	33° 24' S	70° 40' W
33	Buenos Aires	34° 30' S	58° 20' W
34	Caracas	10° 30' N	66° 55' W
35	Quito	00° 15' S	78° 35' W
	Africa		
36	Nairobi	01° 17' S	36° 50' E
37	Timbuktu	16° 49' N	02° 59' W
38	Cairo	30° 03' N	31° 15' E
39	Johannesburg	26°10' S	28°20' E
40	Luanda	08° 50' S	13° 15' E
	Antarctica		
41	McMurdo Sta.	77° 00' S	170°00'E
42	South Pole	90° 00' S	----------

Western Hemisphere

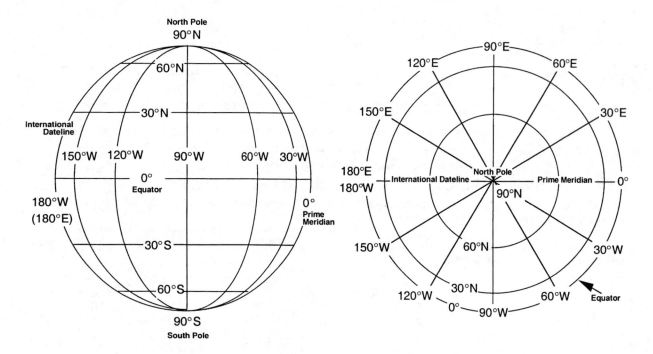

Northern Hemisphere

Eastern Hemisphere

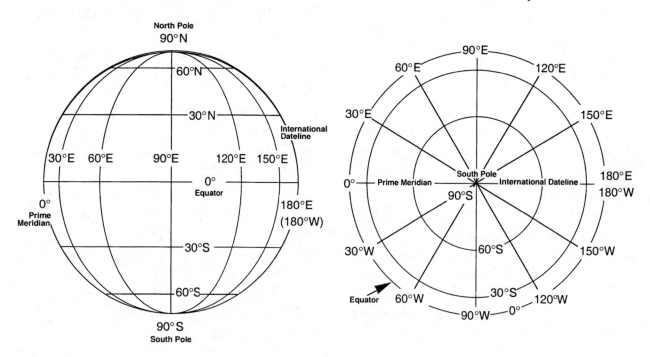

Southern Hemisphere

#2   Navigation

Before *radio navigation systems*, such as Omega or Loran, and before *satellite navigation* with GPS (Global Positioning System), determining your location on Earth or constructing maps was based on **celestial navigation**. Determining ones location north or south of the equator, or *latitude*, had been possible since at least 325 B.C. (Pytheas). However, determining ones position in an east-west reference frame, or *longitude*, was a long-standing problem. Sailors relied on celestial tables or "dead-reckoning".

You can determine **latitude** (position in degrees, minutes, and seconds, north or south of the equator) in the Northern Hemisphere by measuring the angle between the North Star (Polaris) and the horizon. The hand-held instrument used to measure this angle is called a **sextant**. At the North Pole, the North Star is directly above you at 90° (= 90 °north latitude). At the equator, the North Star is on the horizon, or 0° latitude.

You can determine **longitude** (position, in degrees, minutes, and seconds, east or west of the Prime Meridian (0°), which is in Greenwich, England) with a special sea-going clock called a **chronometer**. John Harrison solved the *'longitude problem'* with his intricate and amazingly accurate clocks, 4 in total, built between 1728 and 1760. Captain James Cook (1772-79) made the first accurate maps of Earth's surface (Pacific Ocean) utilizing this new invention.

To determine longitude all you need to know is the time difference between your present location and a reference point, usually Greenwich Mean Time (GMT), where longitude = 0°. The measurement is best done at "local noon," i.e. when the sun is highest in the sky. To determine local noon, observe the path of a shadow made by a vertical shaft, it will be shortest at noon. From this difference you can calculate how much to the east or west you are from your reference location.. Remember, the Earth turns toward the east (e.g., the sun rises in New York before it rises in Chicago).

1). a. Your chronometer is set for Greenwich Mean Time (GMT). High noon at your present location is 5 PM GMT. What is your longitude?

   b. What is your longitude when your chronometer reads 8AM GMT when it is noon where you are?

2). You measure an angle of 21º between the North Star and the horizon with your sextant. High noon at your present location is 10:30 PM GMT. What is your latitude and longitude? Where are you?
   NOTE: Draw a diagram to illustrate how you determined the latitude.

Clearly print your name:

Section:

Student number:

#3	Echo Sounding

Echo sounding can be used at sea to determine water depth. A sound transmitter mounted in the hull of a ship emits high frequency sound waves (many sound waves per second). The *sound waves* travel through the water and are *reflected off the seafloor* because of the different density of the seafloor sediments or rock compared to that of water. These returned "echoes" are detected by a listening device called a hydrophone, which may also be mounted in the hull of the ship. Ships can generate a continuous two-dimensional profile of the seafloor as they travel across the ocean surface. Numerous closely spaced echo sounder profiles (ship-tracks) can be used to construct a **bathymetric map** of the seafloor (a two dimensional representation of the three dimensional relief on the seafloor).

The speed, or velocity, that sound travels through water is approximately **1460 m/sec** (meters per second; ~3275 miles per hour).

1). **Draw a diagram** of sound waves traveling from a ship on the ocean surface through the water, reflected off the seafloor and returning back to the ship.

2). On the basis of your diagram, come up with an equation that may allow you to calculate the water depth below the ship. Remember that you are looking for the depth (**d**), and you know the velocity (**v**) as well as the travel time (**t**) of the sound waves from when they were transmitted until they were received

3). What is the water depth (in meters, **m**) if:

        a. t = 4 seconds?

        b. t = 1 second?

        c. t = 6 seconds?

First 3 letters of last name

#4	Continents and Ocean Basins

A look at any map of the world shows that the Earth's surface is clearly divided into two great regions: those areas that are higher than sea level, the **continents**, and those areas that are covered by water, the **ocean basins**.

1). List one or two ideas that might explain why these two different areas exist on the Earth. You can think of these as **hypotheses**.

The deep structure of the Earth may control the distribution of continents and ocean basins. Your instructor will show the results of an experiment that may help you understand how this process works. You may also refer to p. 94 and 95 to see more about what the interior of the Earth looks like.

2). The first experiment uses materials of the same composition but differing volumes: two blocks of spruce, one is thicker than the other. What are **your observations** about how each floats in water?

3). If the blocks of wood represent continents and ocean basins, what conclusion might you draw about the actual differences between these areas on Earth?

4). In this experiment, what assumptions are you making about the differences between the material of the Earth's crust and that in the Earth's interior?

4). A second experiment uses two different materials (different densities) of the same volume: one block of spruce and one block of oak. **Fill in the table below and calculate density** values for the spruce and oak blocks. What were **your observations** about how each floated in water? Can you make any further modifications or additions to your original answers?

	volume (cm^3)	mass (g)	density (g/cm^3)
spruce block			
oak block			

observations:

Do rocks actually have different densities? Your instructor will show you another experiment to help answer that question.

1). Fill in the table below and **calculate the density of granite** (typical composition of continental crust) **and basalt** (typical composition of oceanic crust) from the last part of the experiment.

	volume (cm^3)	mass (g)	density (g/cm^3)
granite (continental crust)			
basalt (oceanic crust)			

2). Do these results give you any more clues about why there are continents and ocean basins?

Clearly print your name:

Section:

Student number:

#5   Isostasy

       The continental and the oceanic crust plus the upper part of the mantle comprise the **lithosphere**, which rests upon the ductile **asthenosphere**. The total mass of all the material above the asthenosphere, which starts at a depth of about 100 km, is the same everywhere on the Earth. In other words, at this depth a condition of equilibrium, known as **isostasy** or **isostatic equilibrium** is achieved.

See for yourself. On the back of this page is a cross section of the Earth down to the base of the lithosphere 100 km depth. From the information given there, calculate the total mass of material from the Earth's surface to 100 km depth on both the left hand and the right hand side of the diagram. Assume that the shaded columns on each side have a cross-section of 1km^2.

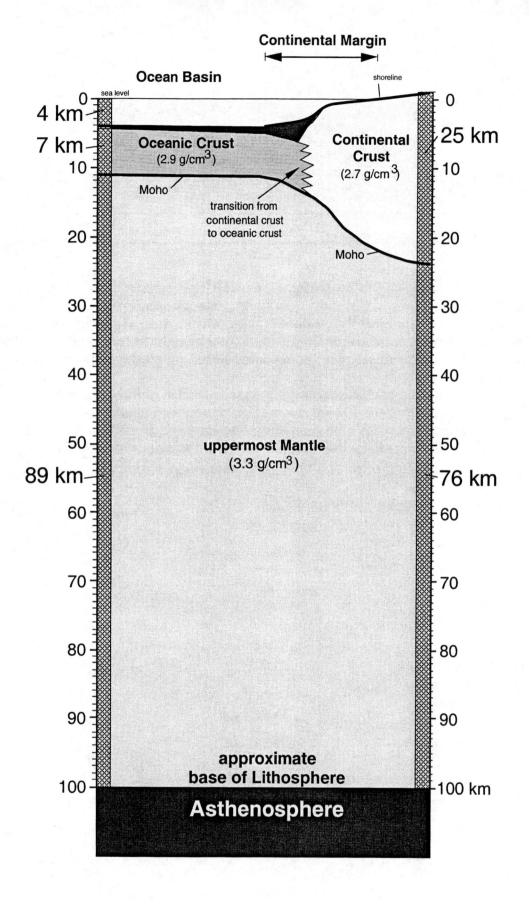

Clearly print your name:

Section:

Student number:

| #6 | Plate Tectonics |

1). The maps on the back of this page illustrate the distribution of global seismicity (earthquake activity) and active volcanoes. Describe any geographic patterns you see and similarities or differences in the distribution of these phenomena.

2) Look carefully at the patterns in and around the Atlantic and Pacific Ocean basins. What are the similarities and differences?

3) Can you come up with any ideas that might explain the patterns you see?

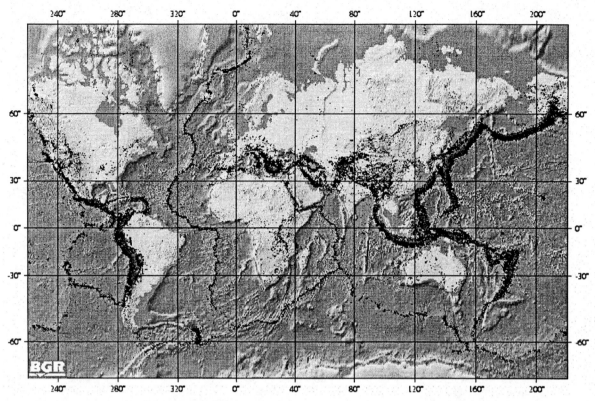

Worldwide earthquakes (seismicity) 1954 to 1998 with magnitude ≥ 4.0
Bundesanstalt für Geowissenschaften und Rohstoffe
http://www-seismo.hannover.bgr.de/wld_seis_deu.htm

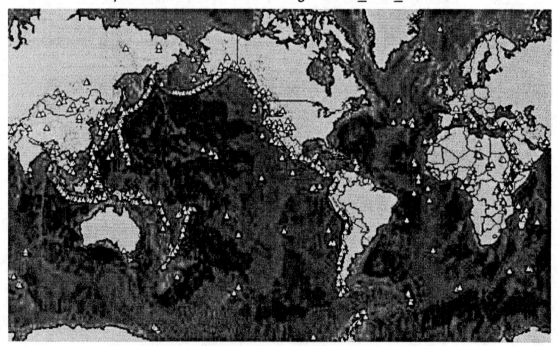

Active volcanoes of the world
Smithsonian Institution Global Volcanism Program
http://www.volcano.si.edu/gvp/volcano/index.htm

Clearly print your name:

Section:

Student number:

First 3 letters of last name

| #7 | Seafloor Spreading |

1). Some areas where earthquakes and volcanic activity often occur together are along **oceanic ridges** where lithospheric plates move apart. If we know the age of the crust and the distance from the ridge center, we can calculate how fast the two plates are moving away from each other. Look at the figures of oceanic ridges (spreading centers) illustrated below:

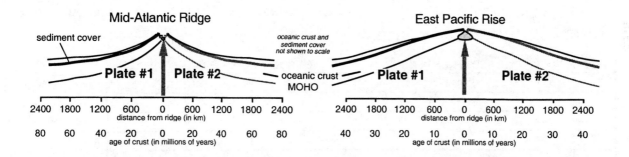

a. Calculate the **half-spreading rate** across the Mid-Atlantic Ridge and East Pacific Rise. This is the rate at which one of the plates is growing.

b. Determine the **full spreading** rate for these same ridges. This is the rate at which the ocean basins are opening.

The map below shows a portion of the North Atlantic Ocean, with the Mid-Atlantic Ridge highlighted. Dashed lines parallel to the ridge show the age of seafloor basalt, in millions of years (my).

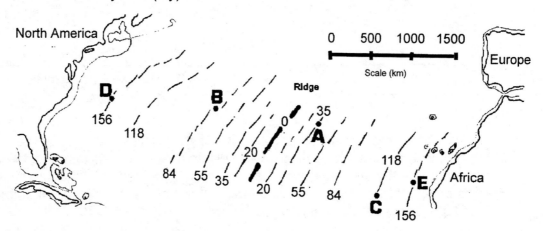

Complete the table for the locations indicated by letter, and answer the accompanying questions

Location	Distance from Ridge	Age of Seafloor	Rate of Movement
	(km)	(my)	(km/my)
A			
B			
C			
D			
E			

3). Is this ocean expanding or contracting? What is the evidence?

4). At what rate is Africa moving away from or toward North America?

5). If the ocean is expanding, when did the continents separate? If it is contracting, when will the continents collide?

16

Clearly print your name:

Section:

Student number:

| #8 | Marine Sediments |

1). What kinds of materials might you expect to find on the bottom of the sea floor? Do you think that there would be a pattern in their distribution?

2). List the major types of marine sediments found in the ocean, and give at least one example of each.

The length of time recorded in sediment layers on the seafloor depends upon the rate of sedimentation of the deposits. Look at a map of the distribution of sediments in the ocean as you answer the following questions.

3).   a. Off the coast of Massachusetts are deposits of terrigenous sediments. If the sedimentation rate here is 10 cm/ 1000 yr., how long did it take to deposit 100 meters of sediment here?

b. Further off shore in the North Atlantic are deposits of ooze, with sedimentation rates of 5 cm/ 1000 yr. How long would it take 100 meters of this sediment to be deposited?

c. Suppose you found a deposit of red clay 100 meters thick in the central Pacific. If the sedimentation rate for the clay is 0.1 cm/ 1000 yr., how long did it take to deposit the 100 meters of sediment?

d. Why do sedimentation rates change in the manner that you have found from the three calculations you have done?

Clearly print your name:

Section:

Student number:

#9	Salinity

1). Ocean water is salty. List some ways you can determine the salt content of sea water.

2). What is the **salinity** of the following sample of sea water?

Mass of sea water sample	630 g
Mass of salt in sample	22 g
Mass of water in sample	

Fill in the missing piece of data in the table. How could you obtain the other data in the table?

3). Where do the salts in the ocean come from? Why aren't the Great Lakes salty?

4). Is the ocean getting saltier with time? How do you know? Explain the reasons.

Clearly print your name:

Section:

Student number:

| #10 | Seawater Chemistry |

1). a. On the accompanying table, list the **five** most common dissolved ions in seawater. Rank them in order from the most common at the top to least common at the bottom. See page 114 in the *Illustrated Guide* section of this book.

   b. In the remaining column, list the five most common ions in rivers. Use the same ranking system as above.

	River Water	Seawater
1		
2		
3		
4		
5		

2). Give some reasons why the lists do not contain the same ions in the same sequence.

Clearly print your name:

Section:

Student number:

#11	Residence Time

1). There are 300 students in this room. If every minute, one student leaves and another arrives, how long does the average student stay in the room?

2). From the concentration data in the first column, determine the mass of Na^+ and Ca^{2+} in the ocean. Then determine the residence time for these two dissolved components.

Mass of the oceans (water + salt) = 14,200 x 10^{20} g

Substance	Concentration		Mass in Ocean	Input Rate	Residence Time
	‰	%	g	g/yr	yr
H_2O	965	96.5		3.8 x 10^{20}	
Na^+	10.5			1.8 x 10^{14}	
Ca^{2+}	0.4			5.2 x 10^{14}	

23

3). How can you explain the difference in the residence time for these components?

Clearly print your name:

Section:

Student number:

#12 Density

1). What is **density**? List some ways that you could measure the density of water.

2). Is salt water more or less dense than fresh water? How could you tell?

3). Design an experiment that would allow you to measure the change in the density of water as temperature changes.

4). Your instructor may show you some experiments about measuring the density of water. Report the results.

Clearly print your name:

Section:

Student number:

#13	The Seasons

1). Why does the Earth experience seasons? Pretend that you're describing this phenomenon to a family member over the telephone (no drawings).

2). Draw a cartoon/picture to depict the cause of the seasons. Would you make any changes or amendments to your original description?

3). Describe your cartoon and explanation to one or two people sitting around you. Would you make any changes or amendments to your original description?

4). Use the space below to write yourself a revised explanation of why the Earth has seasons. Create a diagram or picture to illustrate the major features of this phenomenon.

Clearly print your name:

Section:

Student number:

First 3 letters of last name

| #14 | Solar Angle and Radiation |

1). What is the latitude of our present location?

2). What time(s) of the year does our location experience exactly 12 hours of sunlight per day?

3). What is the sun's angle above the horizon at noon each June 21st? See the diagram on the next page.

4). What is the sun's angle above the horizon at noon each December 21st?

Equatorial view of Earth's orbital plane

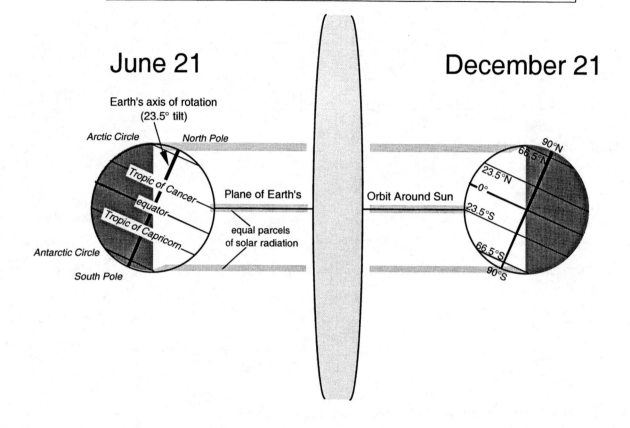

June 21

Earth's axis of rotation
(23.5° tilt)

Arctic Circle

North Pole

Tropic of Cancer

equator

Tropic of Capricorn

Antarctic Circle

South Pole

Plane of Earth's

equal parcels
of solar radiation

Orbit Around Sun

December 21

90°N

66.5°N

23.5°N

0°

23.5°S

66.5°S

90°S

#15	Coriolis Effect

1). On the circle below, place a point in the center and label it NP to represent the North Pole. Now label three additional points: NY (New York City), C (Chicago) and LA (Los Angeles). Near the edge of the circle, draw a curved arrow that represents the direction of Earth's rotation. Label the equator, and put a point "E" on it due south of Chicago.

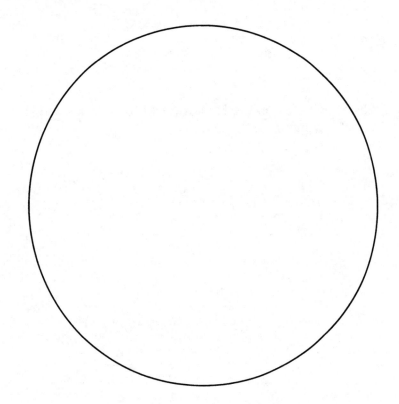

2). In your jet, you take off from the North Pole on a direct heading to Chicago. You place the craft on autopilot and kickback in your seat for a nap. After several hours of flying, are you nearer to Chicago, New York or Los Angeles? Explain in words what is happening.

3). Now you take off from point "E" on the Equator and head due north to Chicago. Will you reach Chicago on this course? Explain.

Clearly print your name:

Section:

Student number:

First 3 letters of last name

| #16 | Winds and Weather |

1). On the globe below, sketch the cloud distribution on as seen on the inside front cover of this book.

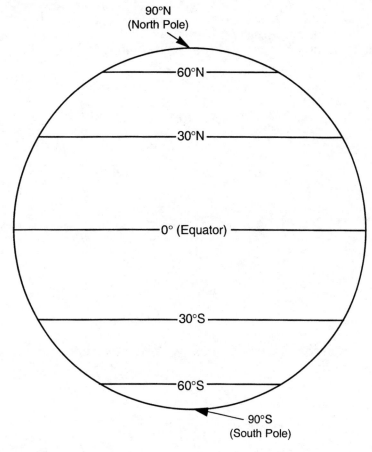

2). Describe in words the distribution of clouds in the photograph. Is it random?

3). In the U.S., what principal direction does our weather come from? Why?

4). Where do Atlantic hurricanes form? What path do they follow as they move northward?

5). Can you explain the reasons why hurricanes follow the paths you have described?

Clearly print your name:

Section:

Student number:

| #17 | Wind-Driven Circulation |

1). Draw the pattern of prevailing winds on the accompanying sketch of the globe by using long arrows.

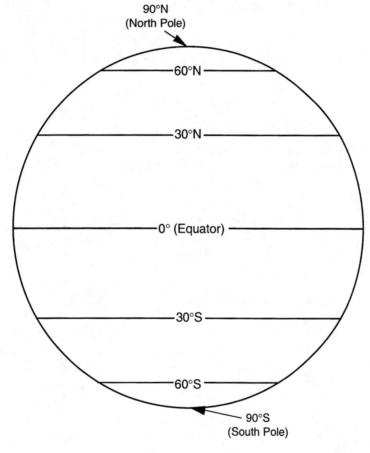

2). Add a shorter arrow depicting the direction of Ekman Transport for each of the original arrows

3). Assume the globe is centered on the Atlantic Ocean. Sketch the outline of the continents on either side. Using bold lines, draw in the resulting surface currents (= **gyres**) driven by the prevailing winds.

Clearly print your name:

Section:

Student number:

#18   Ocean Currents

1). Describe (in words) the major surface current patterns.

2). What direction do the Equatorial Currents flow? Why do they flow in this direction?

3). What happens in the middle of the North Atlantic Ocean?

4). Which way does the current flow off the coast of Chile and Peru? Why is upwelling normally so strong in this area?

Clearly print your name:

Section:

Student number:

#19   Thermocline and Pycnocline

1). Based on what we know about the uneven heating of the Earth, predict the temperature profile of the water column across latitude. Write either "**warm**" or "**cold**" in areas of the diagram below for surface and deep waters of the low, middle and high latitudes.

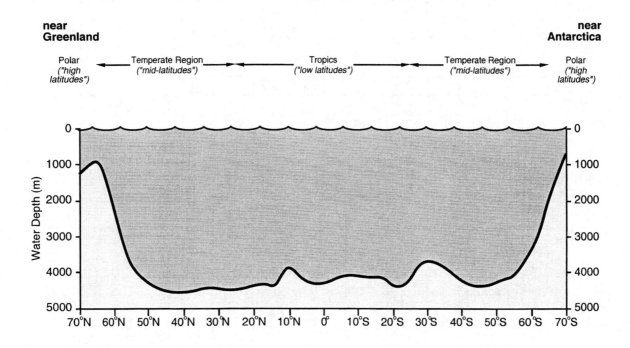

2). Below are temperature and density data collected in the water column at a particular site in the ocean. Fill in the column for σ_t. Plot these data on the graphs below. Label the parts of the curves: **mixed layer, thermocline or pycnocline**, and **deep waters**.

water depth (meters)	temperature (°C)	density (g/cm³)	σ_t
0 (surface)	24	1.023	
100	22	1.0233	
200	18	1.0245	
300	15	1.0252	
400	12	1.0258	
500	10	1.0262	
1000	4	1.027	
1500	3	1.0271	
2000	3	1.0271	
2500	3	1.0271	
3000	3	1.0271	
3500	2.5	1.0273	
4000	2.5	1.0273	

Are these data from low middle or high latitudes?_____

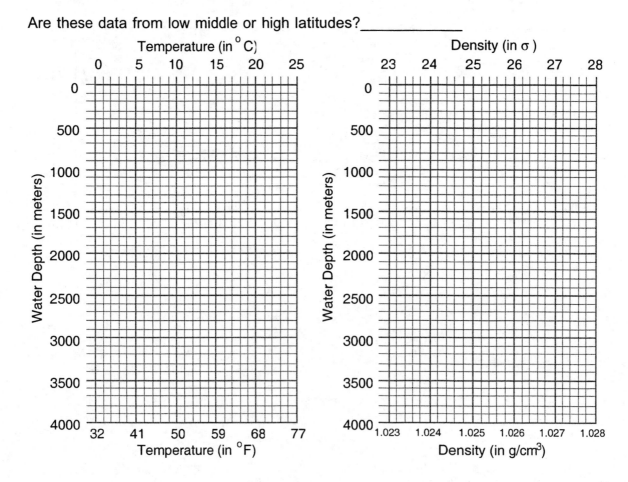

First 3 letters of last name

#20	Waves and wave characteristics

You're out fishing with friends. Your boat is anchored in **90 m of water**.
Fishing is really slow. You're bored, so you decide to apply some of what you learned
in your Introductory Oceanography course to determine the characteristics of the
waves (swell) that are moving the boat up and down. You need to make some
observations. One simple wave characteristic that you can quickly calculate is the
wave period (T). You do this by measuring the **time** it takes successive wave crests
to pass your anchor line. After observing the passage of a dozen wave crests you
determine that **T = 10 sec**.

1). You can calculate the **celerity (C)**(speed) of the waves by knowing only the **wave
period (T)** and using the formula:

   C = 1.56T m/sec

What is the celerity of the swell? **C = _____**

2. Now that you know the celerity of the waves, you can determine the **wavelength (L)**
of the swell by using the formula:

   C = 1.25 √L m

What was the wavelength of the swell? **L = _____**

3). As a check on your calculations, you can determine the **celerity (C)** If you know
both the **wavelength (L)** and **wave period (T)** by using the formula:

   C = L/T m/sec

What is the celerity of the swell? **C = _____**

4). These calculations are true only if the waves are **deep-water waves**, that is, if the **wave base** is shallower than the depth of the ocean through which the waves are moving. Are these deep-water waves at your location?

5). At what water depth do these waves begin behaving more like shallow water waves than deep water waves?

_____m

6). You've determined how fast the waves were traveling in deep water. You can calculate the **celerity** of *shallow water waves* by using the formula:

C = 3.13 $\sqrt{\textbf{d}}$ m/sec   (**d** = depth)

What is the celerity of these waves when water depth = 4 m? ***C*** = _____

What is the celerity of these waves when water depth = 1 m? ***C*** = _____

7). What can you conclude about the celerity (speed) of waves as they approach the shore?

Clearly print your name:

Section:

Student number:

| #21 | Tides |

1). What are tides?

2). What force(s) drive the tides? What observations can we make in order to test this?

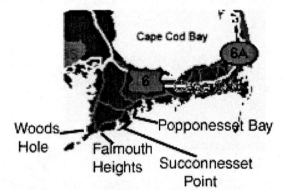

Cape Cod Bay

Woods Hole

Popponesset Bay

Falmouth Heights

Succonnesset Point

3). Examine the tide table below (or other information provided by your instructor). See the map at the left for locations.

TIDES FOR POINTS ON VINEYARD SOUND AND BUZZARDS BAY

DATE	POPPONESSET BAY		SUCCONNESSET POINT		FALMOUTH HEIGHTS		WOODS HOLE	
	High	Low	High	Low	High	Low	High	Low
Nov.28	5:25	11:15	4:16	10:04	3:06	9:16	00:14	7:31
	17:43	23:52	16:34	22:39	15:24	21:51	12:41	20:14
Nov.29	6:27	12:19	5:18	11:06	4:08	10:18	1:15	8:46
	18:47		17:38	23:39	16:28	22:51	13:42	21:24
Nov.30	7:29	00:52	6:20	12:11	5:10	11:23	2:17	10:12
	19:53	13:24	18:44		17:34	23:51	14:43	22:34
Dec. 1	8:30	1:52	7:21	00:39	6:11	12:28	3:16	11:30
	20:57	14:29	19:48	13:16	18:38		15:41	23:35
Dec. 2	9:27	2:51	8:18	1:38	7:08	00:50	4:11	12:32
	21:57	15:31	20:48	14:18	19:38	13:30	16:34	

a. Pick one of the sites. Do high tides occur at the same time every day? What time interval separates successive high tides?

b. Is this interval the same as observed in other towns?

c. Why might the timing of the tides differ so much at the different locations?

4). If the moon is the principal cause for the daily or twice-daily rise and fall of the ocean surface, then what would you predict would be the time between successive tides

for a daily tide?   for a twice-daily tide?

Explain the reason for your answer.

5). The height of high and low tides (the tidal range) changes according to the phases of the moon. Draw a picture to explain what might be causing this.

Clearly print your name:

Section:

Student number:

| #22 | Living in the Ocean |

1). Below is a fish tank which is divided into two parts; a semipermeable membrane separates the two halves. The left side contains fresh water and the right side contains salt water. An osmotic pressure exists across the membrane barrier.

   a). Label the direction that water molecules will move across the membrane.

   b). Fish have body fluids with a salinity of 8 to 15‰. Using the diagram as a guide, explain how fish that live in fresh water and fish that live in salt water maintain their vital water balance.

   c). What would happen to the cells of your stomach if you drank seawater?

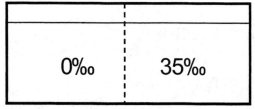

2). Below are two species of copepod (small crustaceans, zooplankton), each is only a couple of millimeters long. Which species is adapted to warm waters? Why?

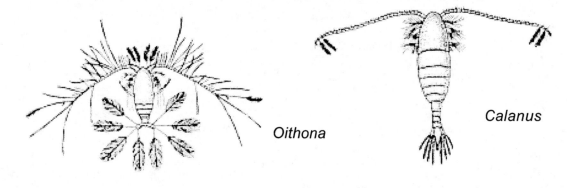

Oithona

Calanus

1 mm

3). Calculate the surface area and volume of the four cubes illustrated below. Next, calculate the ratio of surface area to volume for each. How do these values change with increasing cube size?

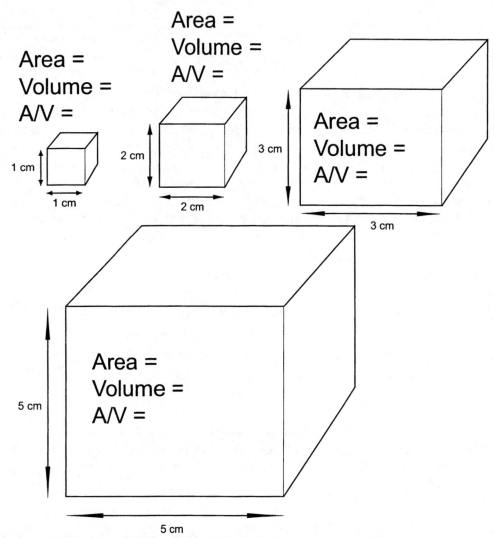

4). Which cube would float best? Why? What does this have to do with marine organisms?

Clearly print your name:

Section:

Student number:

| #23 | Marine Organisms |

1). List 10 organisms that live in the marine environment.

2). How could you classify (or sort) your list of organisms?

2). What physical or chemical properties of seawater might control the distribution of marine organisms? Think about this both in terms of the **abundance** of organisms (= biomass) and the **diversity** of organisms (variety or number of species). In other words, what conditions would favor marine organisms, and what conditions would have a negative effect on marine organisms?

#24   Measuring Productivity

Primary productivity is the amount of new organic matter produced by **photosynthesis (P)** in a given volume of seawater. Since oxygen (O_2) is a by-product of photosynthesis, one way of measuring productivity is to determine the amount of O_2 generated. However, O_2 is also consumed by **respiration (R)**. This can be sorted out by a simple experiment. If pairs of light and dark bottles containing equal amounts of oxygen and an identical population of organisms are placed at various depths in the ocean, then the difference in the O_2 between the light and dark bottles after 24 hours will give a measure of productivity.

Dark bottles = only respiration; O_2 loss occurs. The difference between the amount of O_2 at the start and the amount in the dark bottle at the end of the experiment measures R.

Clear bottles = photosynthesis and respiration during day, only respiration at night. The difference between the amount of O_2 at the start and the amount in the clear bottle at the end of the experiment is influenced by both P and R. This is the **net productivity** (NP = P-R) An increase in O_2 indicates that photosynthesis respiration. When NP = 0, this is the **oxygen compensation level** and the base of the **euphotic zone**.

The difference between the O_2 in the clear bottle and that in the dark bottle shows the total amount of O_2 produced by photosynthesis, even that which was later consumed by respiration. This is the **gross productivity** (GP = NP+R). When GP = 0, this is the base of the **photic zone**.

Here are the data from the experiment :

Depth (m)	1 O_2 at start (ml/l)	2 O_2 in dark bottle (ml/l)	3 O_2 in clear bottle (ml/l)	R	NP	GP
0	6.0	3.8	8.8			
10	6.0	3.9	9.0			
20	6.0	4.0	9.5			
30	6.0	4.0	10.0			
40	6.0	4.1	8.6			
50	6.0	4.3	7.6			
60	6.0	4.4	6.8			
70	6.0	4.7	6.3			
80	6.0	4.9	5.8			
90	6.0	5.1	5.5			
100	6.0	5.3	5.3			

1). How can we determine the **respiration** (R)?.
                 the **net productivity** (NP)?
                 the **gross productivity** (DP)?

2). Plot these values from the table on a graph of O_2 versus depth below

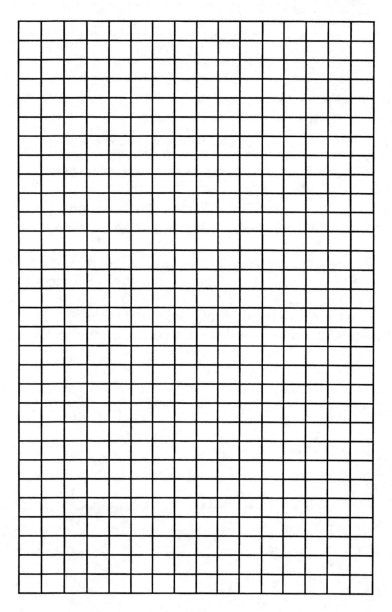

3.) Indicate on the graph, the **oxygen compensation level** and the bottom of the **photic zone**. See pages 152-153 for more information.

Clearly print your name:

Section:

Student number:

#25   The Gulf Stream

The Gulf Stream is a **western boundary** Current that originates in the equatorial Atlantic, flows up the eastern coast of North America and crosses the North Atlantic towards Ireland and England. The water returns to the equatorial region by way of the Canary Current along the western coast of Africa.

1). What causes the Gulf Stream to flow?

2). What are the major physical characteristics of the Gulf Stream? What distinguishes it from the surrounding water?

3). Is productivity high or low within the Gulf Stream? Why?

4). Where is the Sargasso Sea? How did it get that name, and why is it there?

5). What are meanders in the Gulf Stream, and how do they form?

6). How does the Gulf Stream affect the ocean ecology off the coasts of the
     northeastern United States and eastern Canada?

7). How does the Gulf Stream affect the climate of Europe?

Clearly print your name:

Section:

Student number:

First 3 letters of last name

#26 The Intertidal Zone

1). What conditions make the intertidal zone a good place for marine organisms to live?

2). What conditions make the intertidal zone a very difficult place to live?

3. Many organisms develop peculiar ways to adapt to their environment. Organisms in the surf zone need ways to firmly attach themselves to the sea floor so they will not be washed out to sea by the waves. How do these organisms accomplish this?

   a. sea palms

   b. sea urchins

   c. barnacles

   d. sea stars

4. What large marine animals live in the intertidal zone?

5. What are some of the difficulties that organisms encounter during times of low
   tide?

6. How do barnacles eat?   What problems may arise from this type of feeding?

| #27 | Coastal Ecology |

1). What makes coastal environments "fertile"?

2). What is the importance of "seaweed" (= algae) in coastal waters?

3). What are copepods and krill?

4). Give two examples of coastal food chains.

5). How many tons of phytoplankton are needed to feed one Cod during its lifetime?

6). Why are salt marshes important?

7). What are some examples of human impacts on the coastal environment?

#28   The Food Chain

1). What is a food chain (or web)?

2). What kinds of organisms are at the bottom of the chain in the ocean? What kinds are at the top?

3). Where does the energy come from that drives the food chain in the open ocean?

4). How many kg of phytoplankton (primary producers) does it take to grow 1 kg of fish assuming that:
   a. this fish is at the fifth trophic level in the "grazing food chain," and
   b. there is a 10% transfer efficiency between trophic levels.

5). What is the base of the food chain in hydrothermal vent communities?

#29   Benthic Ecology

1). What characteristics do sea stars, sea urchins and sea cucumbers have in common?

2). Where are coral reefs located?

3.) What conditions are necessary for coral reefs to thrive?

5). Describe the main characteristics of coral and name some biological relatives.

#30	Beaches

1). Where does the sand at a beach come from? Does the sand stay in one place? How does it move?

2). What is a barrier beach? Why are there so many of them along the eastern coast of North America?

3). Describe the seasonal changes observed in a beach profile (you may do this with a sketch). What does the summer beach look like? What does the winter beach look like? What causes these seasonal changes?

    a. summer profile:

    b. winter profile:

    c. cause of seasonal changes:

#31   Longshore Drift

1). Describe, in words or a diagram, your understanding of the process of **wave refraction**.

2). In a similar fashion, describe how the process of wave refraction can affect the sand along a shoreline.

3). On the diagrams below, illustrate the answers to the questions about the shorelines.

Use an arrow to show where and in what direction **_longshore currents_** and **_longshore transport_** will occur.

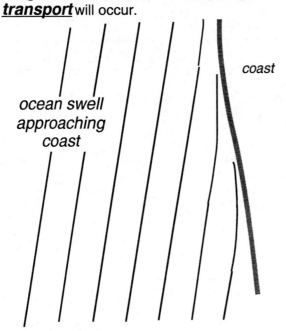

A **_jetty_** is constructed to keep sand out of the harbor. Draw a picture of what the coastline will look like after the jetty is built.

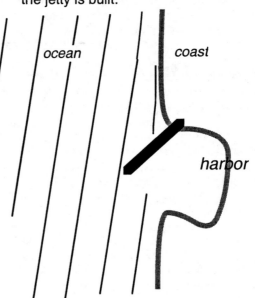

A series of **groins** are built along the coast in an attempt to widen the beach. What happens to the coast after the groins are installed?

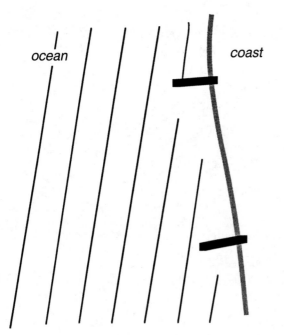

A **_breakwater_** is built a short distance offshore in order to provide safe anchorage for small boats. What happens to the coast after the breakwater is installed?

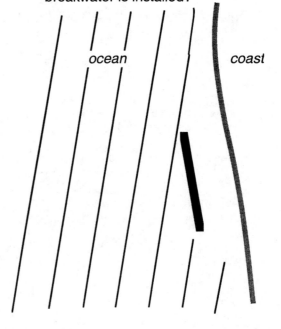

Clearly print your name:

Section:

Student number:

First 3 letters of last name

#32	Estuaries

An estuary is a place where fresh water meets seawater. A common example is the mouth of a river, or the distal end of a river valley that flows to the sea. Using the three profiles below, predict what might happen when river water (0‰) meets saltwater (33‰). Use arrows to depict the flow of each. Using the graph of salinity vs. depth on the right, plot the predicted salinity profile through the oceanward side of the estuary, the middle of this estuary, and the landward side of the estuary. For each diagram use the arrow as an indicator where to draw the graph. Specify the type of estuary you are illustrating._____

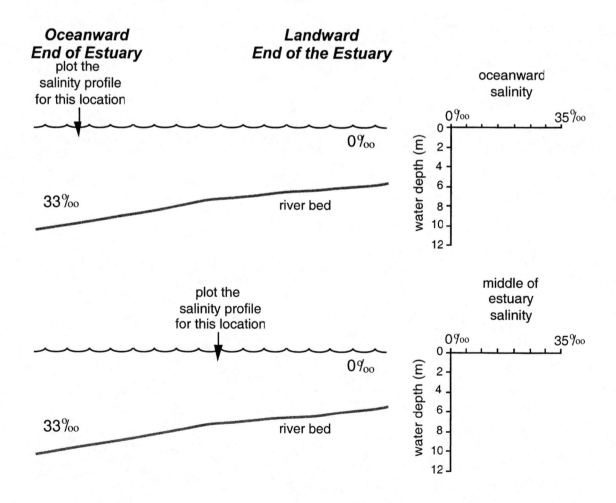

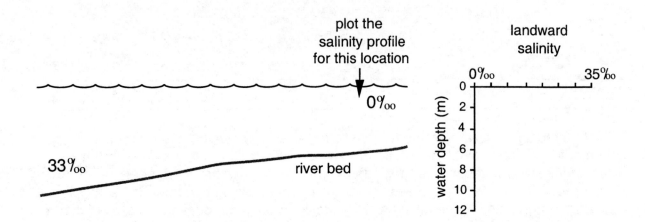

Clearly print your name:

Section:

Student number:

#33   Coastal Erosion

During more than 350 years of European settlement, the coast of Cape Cod has been inundated by strong storms 84 times! During the winter of 1978, and again during the "Perfect Storm" of 1991 when powerful nor'easters pounded the coast. The effects were widespread and devastating, but out of sight and out of mind of many "summer people".

1). The coast of Cape Cod is retreating landward. Why is this happening?

2). Sand dunes are a common feature on Cape Cod and other barrier beaches. Where are dunes usually located with respect to the beach? Can they be of value in preventing beach erosion?

Minute paper. An increasingly large population in North America and the world is living directly on the coast. Many communities are being damaged time and again by storms as well as by slower processes of coastal erosion. Are there effective measures against coastal erosion? What can be done to address the problem? In the space below, write down your ideas on some remedies that may work and others that are likely to fail.

Clearly print your name:

Section:

Student number:

#34   The Ocean and Global Environment

1). At the latitude of Massachusetts, the temperature at the coast is warmer in the winter and cooler in the summer than the temperature in the continental interior. Can you come up with an explanation?

2). What role does the ocean play in global warming?

69

3). Look at the false-color satellite image of the oceans shown on the front cover of this book. Describe what you see (colors, patterns, notable features). Can you interpret what it might mean?

Clearly print your name:

Section:

Student number:

#35   Overfishing

1). When did fishing become an environmental problem?

2). Does overfishing have economic impacts?

3). Who has jurisdiction over fish resources on the high seas? What problems does this cause?

4). What is a solution to the problem of overfishing?

Clearly print your name:

Section:

Student number:

First 3 letters of last name

| #36 | Marine Pollution |

1). List 10 things that can pollute the ocean environment. Indicate by numbering which three you think are the greatest problems (#1, #2, #3)

2). Explain the reasons behind your choice of the three greatest marine pollution problems.

3). What causes Minimata disease? How did it get its name?

4). What is a "red tide" and how does it affect biological productivity in the area around it?

5). How can a salt marsh be used to aid in sewage treatment?

Clearly print your name:

Section:

Student number:

#37	Global Warming

Greenhouse gases such as carbon dioxide (CO_2) and methane (CH_4) trap long-wavelength solar radiation in the atmosphere. There is strong evidence that average global temperatures are slowly rising. A likely cause of global warming is the clear trend of increasing concentrations of greenhouse gases due to the burning of fossil fuels. The plot below shows the monthly mean concentration of as measured near the summit of Mauna Loa in Hawaii since 1958 (the International Geophysical Year).

1). Describe the trend of CO_2 over the last 40 years. Has CO_2 risen at a constant rate?

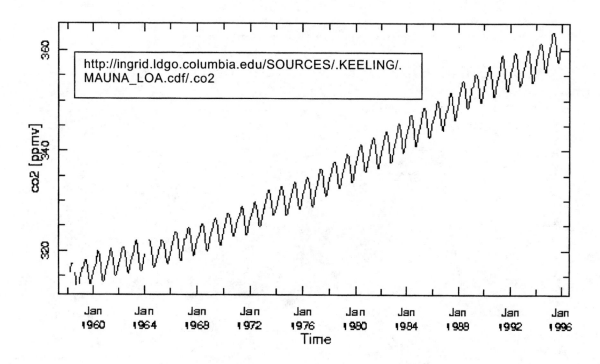

The plot below shows a trend measured from an ice core collected in Antarctica.

- The **concentration of CO_2** has varied from relatively high values during <u>interglacial</u> times such as today and ~140,000 years ago, and relatively low values during <u>glacial</u> times such as 20,000 and 160,000 years ago.
- During glacial times, the concentration of CO_2 in the atmosphere is low. During interglacial times like today, average global temperatures are warmer and atmospheric CO_2 concentrations are greater.

3). What is the **natural range of variability in the CO_2** prior to growth of the human population and industrialization?

4.) Label on the graph where today's concentration of CO_2 would be.

5). Indicate the times when the Earth was colder (glaciation) and warmer (interglacials). What would sea level be like during these times? Why?

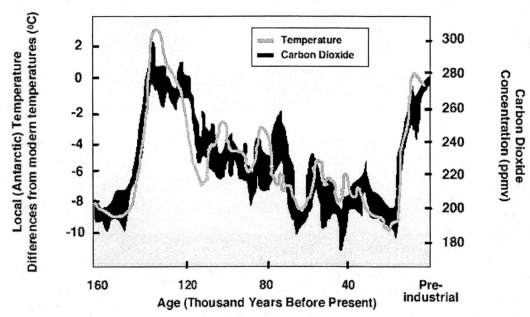

Local Temperature Change and CO_2 Concentrations Over the Past 160,000 Years

Derived from Antarctic ice cores

Source: Based on IPCC (1990)

♻EPA

United States Environmental Protection Agency

76

Clearly print your name:

Section:

Student number:

A

Clearly print your name:

Section:

Student number:

First 3 letters of last name

B

Clearly print your name:

Section:

Student number:

C

Clearly print your name:

Section:

Student number:

First 3 letters of last name

D

Clearly print your name:

Section:

Student number:

First 3 letters of last name

E

Part 2: *Illustrated Guide*

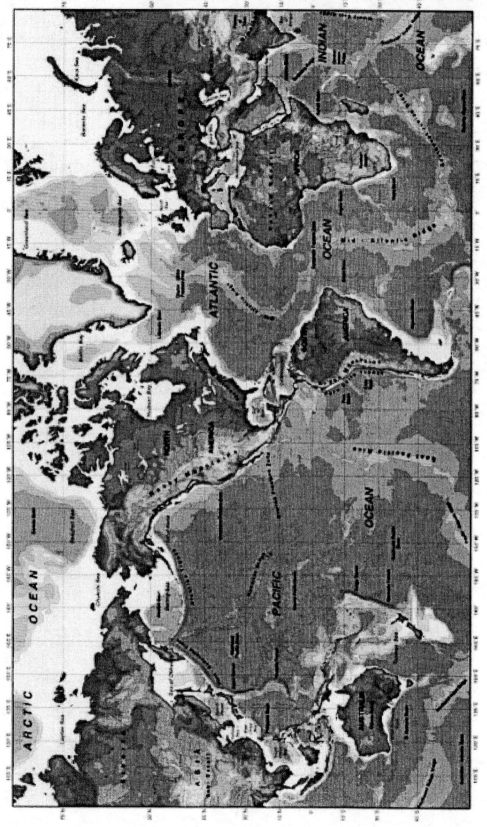

Latitude & Longitude

Lines of **latitude** and **longitude** represent a grid system used to precisely describe locations on the Earth's surface. These reference lines are based on angular relationships to Earth's center. Both latitude and longitude are expressed in degrees (°); each degree is divided into 60 minutes ('), and each minute is divided into 60 seconds (").

Latitude Terminology
- "parallels" = east-west grid lines that are parallel to the equator
- 0° at the **Equator**, 90°N at the **North Pole**, and 90°S at the **South Pole**
- 0°-90°N = **Northern Hemisphere**
- 0°-90°S = **Southern Hemisphere**
- latitudinal distances do not vary anywhere on Earth's surface, at all locations:
    - 1° latitude = 111.32 km (= 60 nautical miles)
    - 1' latitude = 1 nautical mile

Longitude Terminology
- "meridians" = north-south grid lines that intersect at the poles
- 0° = **"Prime Meridian"** (passes through Greenwich, England)
- 180° = halfway around Earth (roughly approximates the international dateline)
- 0°-180° to the west of Greenwich is the **Western Hemisphere**
- 0°-180° to the east of Greenwich is the **Eastern Hemisphere**
- longitudinal distances vary with distance from the equator:
    - at 0° latitude, 1° longitude = 111.32 km (= 60 nautical miles)
    - at 30° latitude, 1° longitude = 96.49 km (= 52 nautical miles)
    - at 60° latitude, 1° longitude = 55.80 km (= 30 nautical miles)
    - at 90° latitude, 1° longitude = 0 km

1 nautical mile = 1.15 statute mile = 1.85 km
1 km = 0.62 statute mile = 0.54 nautical mile

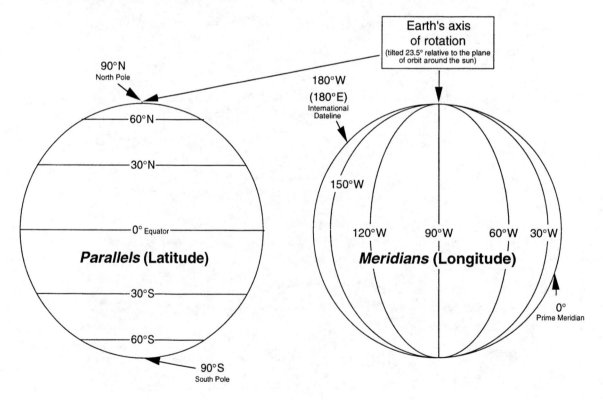

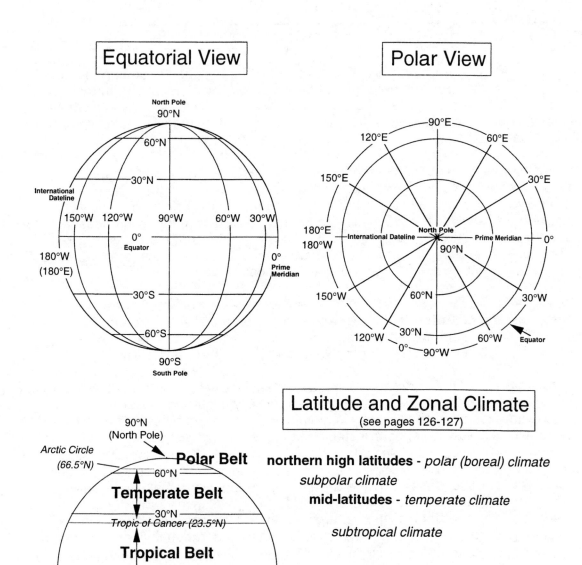

Latitude and Zonal Climate
(see pages 126-127)

northern high latitudes - *polar (boreal) climate*

subpolar climate

mid-latitudes - *temperate climate*

subtropical climate

low latitudes - *tropical climate*

subtropical climate

mid-latitudes - *temperate climate*

subpolar climate

southern high latitudes - *polar (austral) climate*

Because Earth's axis of rotation is tilted at a 23.5° angle relative to its plane of orbit around the sun, we experience seasonal changes in solar radiation at all latitudes (see p. 118-119). For example:

Equator (0°) - the sun is directly overhead on or about March 21 & September 21

Tropic of Cancer (23.5°N) - the sun is directly overhead on or about June 21

Tropic of Capricorn (23.5°S) - the sun is directly overhead on or about December 21

Arctic Circle (66.5°N) - the sun stays above the horizon all day on or about June 21, and doesn't rise at all on or about December 21

Antarctic Circle (66.5°S) - the sun stays above the horizon all day on or about December 21, and doesn't rise at all on or about June 21

The Earth as a Habitable Planet

Seawater covers nearly 71% of the Earth's surface. The Earth is unique among the nine planets in our solar system because of temperatures that allow water to occur in the liquid state. In fact, Earth is the only planet where water is known to occur naturally in all three states: gaseous, liquid, and vapor. It is the presence of water that moderates the extremes of climate and sustains life.

The Earth functions as a series of interconnected "systems", including the Geosphere, Biosphere, Atmosphere, Hydrosphere, and Cryosphere (see facing page). The Hydrosphere plays a fundamental role in the processes that make Earth habitable. For example, the rain and snow that sustains life on land is derived from the world ocean. Earth's interconnected systems are linked by the **flow of water** and the **recycling of carbon**. The linkages include the effects of volcanism and plate tectonics, the hydrologic cycle and climate, as well as life itself and biogeochemical cycling.

Earth systems are driven by two power plants: an external source of energy and an internal source. **Solar radiation** is the external source, responsible for heating the atmosphere and ocean, generating the winds and ocean currents, creating our weather and the erosion of the land, and powering photosynthesis and the webs of life.

Internal energy is the result of **radioactive decay**. Many elements, such as hydrogen (H), carbon (C), and oxygen (O), have two or more naturally occurring **isotopes***. Isotopes of the same element differ slightly in atomic mass, or weight, because of differing numbers of **neutrons**. Most isotopes are stable but some are not. For example, the element uranium (U) has two unstable isotopes (U^{235} and U^{238}). These unstable isotopes spontaneously change to stable isotopes of a different element, in this case lead (Pb), in a process called **radioactive decay**. There are three principal ways that a **radioactive isotope** can spontaneously decay to a more stable form, but all involve the generation of **heat** and the emission of **radiation** (high-energy subatomic particles such as alpha or beta particles, or gamma rays). Some radioactive isotopes undergo many decay steps on their way to becoming stable, such as $U^{235} \rightarrow Pb^{207}$ and $U^{238} \rightarrow Pb^{206}$. Because each type of radiogenic isotope has a unique rate and mode (pathway) of decay, the ratio of radioactive parent isotopes to stable daughter isotopes in the minerals that make up certain types of rocks can be used to calculate **geologic time**.

Over geologic time, Earth has been slowly losing this heat of radioactive decay. Heat is transferred from deep within the Earth to the surface by way of **conduction** (transfer of heat from one atom to another) and **convection** (differential heat loss that causes movement of mass). Volcanoes spew out gases, including water and carbon dioxide, while subduction of crustal rocks, sediments, and water back into the mantle help to perpetuate an ever-changing landscape, atmospheric composition, climate, and biodiversity.

The Earth is *dynamic*; it's exterior is constantly being reshaped by the processes of plate tectonics, the hydrologic cycle, and biotic evolution. The ocean plays a major role in maintaining a livable environment on our home planet.

*The nucleus of all atoms contains **protons** (positively charged particles) and **neutrons** (particles with no charge); both of these subatomic particles have nearly identical masses. It is the number of protons in the nucleus of an atom (**atomic number**) that identifies what kind of element it is and determines the unique chemical properties and characteristics of that element, while it is the number of neutrons that determines how many different isotopes an element has.

the Earth is *dynamic*, ever-changing;
it functions as series of interconnected "systems"
driven by the *internal energy* of radioactive decay,
conductive and convective heat loss,
and by the *external energy* of solar radiation

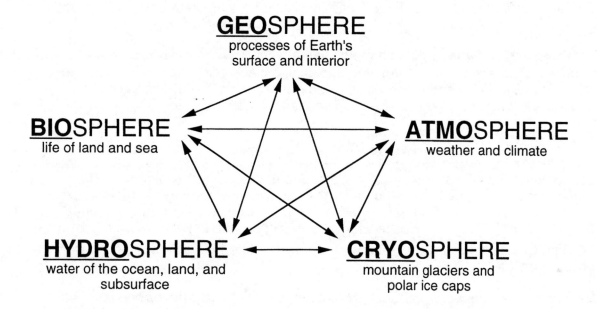

GEOSPHERE
processes of Earth's
surface and interior

BIOSPHERE
life of land and sea

ATMOSPHERE
weather and climate

HYDROSPHERE
water of the ocean, land, and
subsurface

CRYOSPHERE
mountain glaciers and
polar ice caps

**Earth's interconnected systems are linked
by the <u>flow of water</u> and the <u>recycling of carbon</u>:**
- plate tectonics and degassing
- hydrologic cycle and climate
- life and biogeochemical cycles
- surficial and oceanic processes

the ocean covers nearly 71% of Earth's surface
and is an integral part of Earth System Science

History of the Earth

Our **Solar System** formed ~4.6 billion years ago based on radiometric age dates of meteorites collected on the Earth's surface, and from lunar rocks collected during the Apollo missions of the early 1970's. Early Earth history (4.6-4.0 billion years ago) was characterized by **very high heat flow** due to gravitational contraction and radiogenic decay, and by intense **meteorite and comet bombardment**. During this time the Earth segregated into **concentric spheres of differing density** (see p. 94-95): 1) iron-nickel **core**, 2) dense silicate **mantle**, 3) less dense silicate **crust**, and 4) gaseous **atmosphere**. As Earth's surface began to cool, the **crust formed**. The oldest preserved crustal rocks are ~4.0 billion years old. Extensive **volcanism and outgassing** from the mantle produced an early atmosphere rich in H_2O, CO_2 and nitrogen. As Earth's surface cooled further, *water condensed* and the **ocean formed**. Some of Earth's surface water may have come from comets, which are composed of ice and rock.

Gases emitted by volcanoes contain molecules of CO_2, H_2O, and HCl. The carbon dioxide and water can combine to form carbonic acid, a weak acid. Acid rain accelerates the process of **chemical weathering** of exposed crustal rocks. The dissolved ions ("salts") in seawater are derived from volcanic gases and the dissolution of rocks.

$$CO_2 + H_2O \Rightarrow H_2CO_3$$
in atmosphere   carbonic acid (a weak acid)

Some elements or substances occur in the atmosphere, ocean, and sediments in far greater amounts than can be accounted for by the chemical weathering of crustal rocks. These substances are called "**excess volatiles**" and they originated from the mantle via volcanism (CO_2, H_2O, Cl, N, S). Notice what's here: carbon (the building block of life), **water**, and **chlorine** (chloride Cl^- is the most common solute in seawater); notice what's not: **free oxygen** (O_2).

The earliest forms of life were probably **anaerobic chemosynthetic bacteria**. These **prokaryotic microbes** (cells without a nucleus or organelles; see p. 146) lived in world lacking free oxygen. They manufactured their own food using energy from *chemical oxidation*. Photosynthetic **cyanobacteria** ("blue-green algae") appeared by ~3500 Ma. **Photosynthetic autotrophs**, including cyanobacteria, are organisms that manufacture organic compounds (carbohydrates, proteins, lipids) using energy from *sunlight*. Oxygen is a by-product of **photosynthesis**:

$$CO_2 + H_2O + \textbf{inorganic nutrients} + \text{solar radiation} \Rightarrow \textbf{organic compounds} + O_2$$

Microbial mats of cyanobacteria became widespread in shallow marine waters during the later Archean and Proterozoic Eons. These organisms were responsible for altering Earth's atmosphere and making it possible for the evolution of multicellular life. The evolution of our present atmosphere took many hundreds of millions of years (3500-543 Ma), and involved the gradual build-up of free O_2, draw-down of CO_2, and development of the ozone layer which protects organisms from harmful ultraviolet solar radiation. Single-celled **eukaryotic protists** (cells with a nucleus and organelles) evolved via bacterial **symbiosis** early in the Proterozoic Eon and multicellular animals evolved late in the Proterozoic. The "Cambrian explosion" marks the beginning of the Phanerozoic Eon and represents the rapid diversification of **invertebrate animals** with shelly hard parts 543 million years ago. This event was triggered in part by a critical threshold in atmospheric oxygen concentration. The great diversification of animal and plant life since 543 Ma has also included catastrophic extinction events called "mass extinctions".

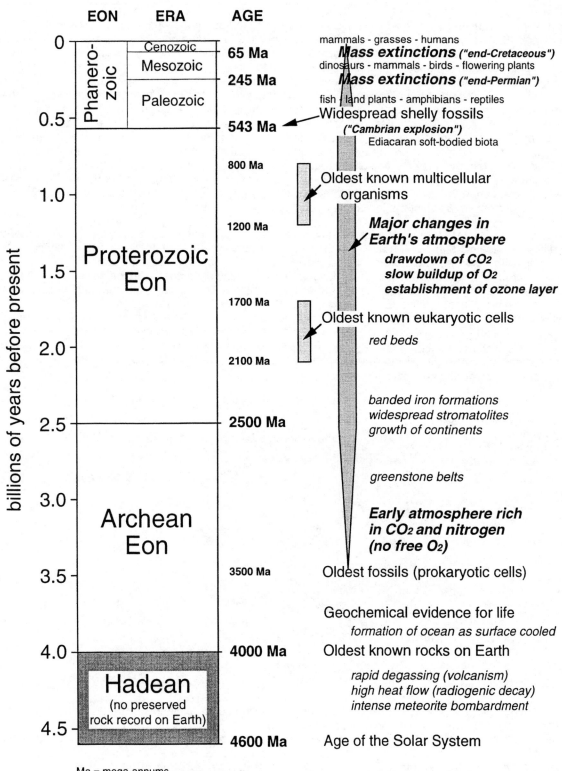

Ma = mega-annums
millions of years before present

Earthquakes & Seismic Waves: Earth Structure Revealed

Earthquakes are caused by the sudden release of energy locked-up by frictional forces between solid bodies of rock. Earthquakes are commonly caused by slip along a fault zone or by volcanic activity. Rocks accumulate the strain induced by stresses of solid materials being pushed and pulled in different directions in the Earth's Crust and upper Mantle. The strain accumulates to the point of catastrophic failure at which time the energy is released as **seismic waves**; what we call an earthquake. Seismic energy radiates from its point of origin, called the **focus**, and travels through the Earth. The point on the Earth's surface directly above the focus is called the **epicenter**. This energy is transmitted in multiple forms. **Rayleigh waves** transmit energy along the surface similar to water waves and are primarily responsible for the damage produced near the epicenter. Primary or pressure waves, called **P-waves**, migrate through solid and liquid materials in an alternating push-pull, or compression-relaxation, motion similar to that of sound waves. Secondary or shear waves, called **S-waves**, migrate through solid materials in a transverse motion analogous to a wave initiated by whipping a rope or a rug. S-waves cannot travel through liquids. P-waves propagate nearly twice as fast as S-waves. By comparing the difference in **arrival times** of the P- and S-waves at numerous locations around the Earth, it is possible to pinpoint the focus and epicenter of an earthquake.

As P- and S-waves migrate through the Earth, they encounter materials of differing composition and **density**. These differences cause some of the energy to be **reflected** off the surface of change. The remainder of the energy is **refracted** (bent) due to the change in physical properties. Both P- and S-waves travel faster through denser rocks. Seismic refraction is analogous to the bending of light reflected off a spoon in a clear glass of water. Studying the arrival times of seismic waves at multiple recording stations around the globe has revealed the layered structure of the Earth's interior. These concentric layers are arranged according to density, with the densest materials making up the Core (iron and nickel) and the least dense materials making up the Crust (felsic and mafic silicate rocks). **Density** is the amount of mass per unit volume, measured in grams per cubic centimeter (g/cm^3). *Think of density as the amount of clothes (mass) contained in a suitcase (volume); compare, for example, the weight of a suitcase containing a few articles of clothing (less dense) to the same suitcase stuffed with clothes (more dense).*

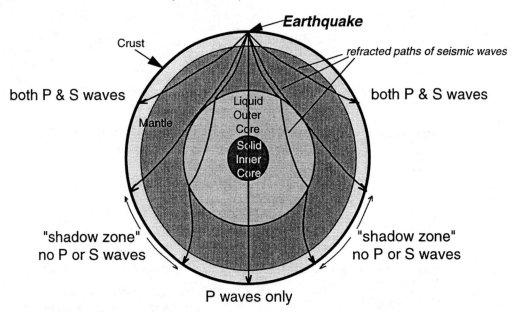

There are several steps of greater seismic velocity with increasing depth within the Earth. Prominent among these are the marked increases of seismic velocity at the Crust/Mantle and Mantle/Core boundaries. Such changes in velocity are attributed to changes in the composition and/or physical properties of the rocks due to changing conditions of temperature, pressure, and volatile content (e.g., water) with depth. The increase in seismic velocity at the Crust/Mantle boundary is called the **Mohorovicic Discontinuity**, or **Moho**, and it marks the transition to rocks of different composition and greater density. A zone of lower seismic velocity in the upper Mantle corresponds to an interval of easily deformed rocks called the **Asthenosphere**. The **D" Layer** marks the transition from mantle rocks rich in silicon to the core composed of iron and nickel. We know that part of the core must be liquid because no S waves are recorded on the side of the Earth opposite an earthquake, and we know that the core must be very dense because P-waves are highly refracted thereby creating a broad **shadow zone** where no P or S waves are recorded.

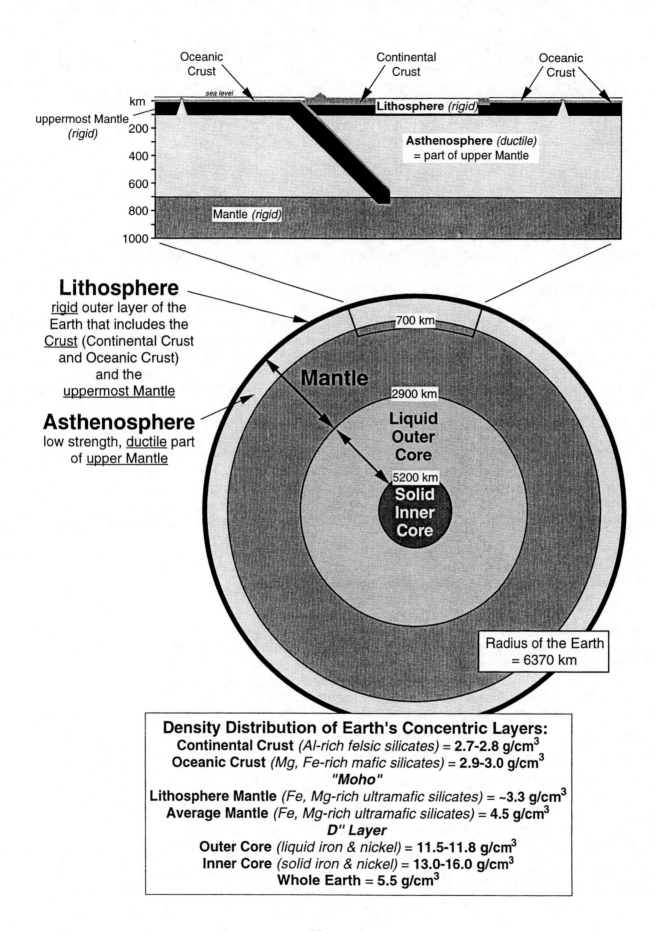

Oceanic Crust

Continental Crust

Oceanic Crust

sea level

km

Lithosphere *(rigid)*

uppermost Mantle *(rigid)*

200

400

600

800

1000

Asthenosphere *(ductile)*
= part of upper Mantle

Mantle *(rigid)*

Lithosphere
rigid outer layer of the
Earth that includes the
Crust (Continental Crust
and Oceanic Crust)
and the
uppermost Mantle

Asthenosphere
low strength, ductile part
of upper Mantle

700 km

Mantle

2900 km

Liquid Outer Core

5200 km

Solid Inner Core

Radius of the Earth
= 6370 km

Density Distribution of Earth's Concentric Layers:
Continental Crust *(Al-rich felsic silicates)* = **2.7-2.8 g/cm^3**
Oceanic Crust *(Mg, Fe-rich mafic silicates)* = **2.9-3.0 g/cm^3**
"Moho"
Lithosphere Mantle *(Fe, Mg-rich ultramafic silicates)* = **~3.3 g/cm^3**
Average Mantle *(Fe, Mg-rich ultramafic silicates)* = **4.5 g/cm^3**
D" Layer
Outer Core *(liquid iron & nickel)* = **11.5-11.8 g/cm^3**
Inner Core *(solid iron & nickel)* = **13.0-16.0 g/cm^3**
Whole Earth = 5.5 g/cm^3

Plate Tectonics I: Lithosphere & Asthenosphere

Despite the marked increase in seismic wave velocity across the Crust/Mantle transition (**Moho**), the uppermost Mantle behaves like the Crust; the Crust and uppermost Mantle are *rigid*. This rigid outer shell of the Earth is called the **Lithosphere**. The base of the Lithosphere is at a depth of approximately 100 km, although this depth varies.

The region of the upper Mantle underlying the Lithosphere behaves like a *ductile* substance. In other words, it is capable of *flow* like hot plastic or hot asphalt, although it is not liquid. This low strength region of the upper mantle is called the **Asthenosphere** (see p. 94). This part of the Mantle is also referred to as the **low velocity zone**, attesting to its properties, which differ from the rest of the Mantle. The base of the Asthenosphere is approximately 700 km, although some geophysicists place the base closer to 350 km.

The core of the Earth is very hot (>5000°C or 9000°F) accounting for the liquid nature of the outer core. The solid inner core has been slowly growing at the expense of the liquid part of the core as the Earth has gradually cooled over time. The Earth has been slowly losing heat by **conduction** (transfer of heat from one atom to another) and **convection** (differential heat loss that causes movement of mass) throughout its history. This **heat flow** deep within the Earth's core and mantle creates **thermal gradients** within the Asthenosphere. Because of its weaker physical properties, these thermal gradients drive **convective motion** in the Asthenosphere; essentially a conveyor-like flow of hot, easily deformed rocks. Convection in the Asthenosphere creates stresses in the overlying rigid Lithosphere due to the upward, lateral, and downward flow of rock. **Upwellings** are sites at the surface where old continents are rifted apart, oceanic crust is produced, and new ocean basins are formed, while **downwellings** are sites where old oceanic crust is recycled back into the mantle, volcanic islands and volcanic mountain ranges are formed, or where continents collide. Because of these deep-seated stresses, the outer shell of the Earth is broken into numerous rigid **Lithospheric Plates** (also called tectonic plates).

The <u>rigid plates interact</u> with each other in one of three ways:
  1. they collide (convergence)
  2. they move apart (divergence), and
  3. they slide past one another (strike-slip motion)

Plate tectonics is a theory about global dynamics that explains the relationship between the rigid exterior of the Earth, the Lithosphere, and the differential heat flow from the Mantle, particularly the convective motion of the Asthenosphere. The interaction of lithospheric plates is substantiated by the distribution and processes of volcanism, seismic activity, and mountain building. Plate tectonics grew out of the concepts of **continental drift** and **seafloor spreading**.

Continental drift was a hypothesis championed by Alfred Wegener early in the 20th century to explain, among other geological evidence, the jigsaw puzzle fit of the continents such as South America and Africa. Wegener proposed that the continents were once one large landmass, which he called **Pangea**. According to Wegener's hypothesis, Pangea split up into the present continents, which then drifted over the denser rocks of the ocean basins. We now know that the <u>continents **move with** the ocean crust</u> as large lithospheric plates. *Seafloor spreading* was a hypothesis proposed by Harry Hess in 1960. Hess proposed that ocean crust is produced at mid-ocean ridges and then moves symmetrically away from the ridge as new material is added in the central rift valley. In this view, the continents are pushed away from the mid-ocean ridges and toward the deep-sea trenches where oceanic crust is returned to the Mantle and partially melted as part of the Asthenosphere's conveyor-like motion. Paleomagnetic studies of ocean crust during the early 60's (F. Vine, D.H. Matthews, and L.W. Morley) largely demonstrated the validity of the seafloor spreading hypothesis.

Lithosphere
rigid outer shell of the Earth;
includes the Crust and
uppermost Mantle

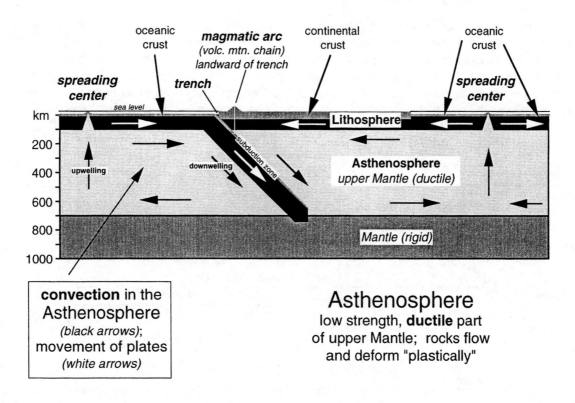

Asthenosphere
low strength, **ductile** part
of upper Mantle; rocks flow
and deform "plastically"

convection in the Asthenosphere *(black arrows)*; movement of plates *(white arrows)*

3 types of plate boundaries:
*all are associated with **seismic activity (earthquakes)***

1. **convergent boundaries** - 2 plates collide, the denser of the two is subducted beneath the other (*=downwelling*); zone of *subduction* is marked by a deep-sea **trench**, <u>volcanism</u> on the over-riding plate, and shallow-, intermediate-, and deep-focus <u>earthquakes</u> along the slope of the subducting slab (**subduction zone**)

2. **divergent boundaries** - 2 plates move apart as new oceanic crust is produced at **spreading centers** by material introduced from the Mantle (*=upwelling*), marked by <u>volcanism</u> in/near the **central rift valley** and shallow-focus <u>earthquakes</u>

3. **strike-slip boundaries** - 2 plates slide past one another along **transform faults**, marked by shallow-focus <u>earthquakes</u>

Isostasy: Continents & Ocean Basins

The Earth's surface has two distinct levels: the **continents** and the **ocean basins**.
- the <u>average height of the continents</u> is 840 m or 0.84 km above sea level (=2760 ft. or 0.5 mile)
- the <u>average depth of the ocean basins</u> is 3800 m or 3.8 km below sea level (=12,500 ft. or 2.4 miles)

The Lithosphere is not homogeneous because there are two types of Crust. Differences in crustal thickness and density are responsible for the two prominent levels on the Earth's surface. **Continental Crust** is thicker and *less dense* (more buoyant) than oceanic crust. Therefore, the continents stand higher than the adjacent ocean basins. In addition, the continents are preferentially preserved through geologic time because they are too thick and buoyant to be subducted at convergent margins. **Oceanic Crust** is thinner and *more dense* than continental crust, and as a consequence, oceanic crust is preferentially subducted during plate collisions.

Oceanic Crust	Continental Crust
<u>thin, more dense</u> • composed of dark-colored *mafic* rocks like **basalt** (rich in Si, Mg, & Fe) • avg. density: **2.9-3.0 g/cm^3** • thickness: **4-10 km**	<u>thick, less dense</u> • composed of light-colored *felsic* rocks like **granite** (rich in Si & Al) • avg. density: **2.7-2.8 g/cm^3** • thickness: **25-40 km**
forms ocean basins	"buoyant" continents stand high
subducted at trenches during plate convergence	**preferentially preserved** during plate convergence
ocean basins <200 million years old *(i.e., the present ocean basins are relatively **young** features)*	continents >3500 million years old *(i.e., the continents are **old**)*

The **ductile Asthenosphere** supports the **rigid Lithosphere** (see p. 96-97). A balance, or condition of equilibrium, is maintained between crustal blocks of different densities such as oceanic crust (more dense) and continental crust (less dense); this is called **isostasy**.

The Asthenosphere and Lithosphere can accommodate changes in the *redistribution of load* that occur in the crust by bending and flexing. For example, mass added to the crust by the growth of an ice sheet, or a volcano, or an island arc, or mountain range, will depress the crust. Later, the crust will rebound if the *load is removed* due to the retreat of the ice sheet or erosion of the mountains. In other words, the crust and Lithosphere maintain "**isostatic equilibrium**" with the underlying Asthenosphere.

*The profile through the lithosphere at the right shows a **"passive" continental margin** created by divergent plate motion. The ocean crust and continental crust are part of a single lithospheric plate (no plate boundaries are shown in this diagram). Notice that there is no subduction of oceanic crust and no volcanism. The transition between continental crust and oceanic crust was created when an old continent split (rifted) into two. A "twin" passive continental margin would have been formed on the opposite side of the ocean basin shown here. The process of **seafloor spreading** has subsequently moved the twin (conjugate) margins apart as the ocean basin widened over time. By contrast, an **"active" continental margin** is created when oceanic lithosphere is subducted beneath continental lithosphere thereby resulting in <u>volcanism</u> on the overriding plate and <u>frequent seismic activity</u>. In this case, a <u>deep-sea trench</u> would mark the boundary between the plates (see p. 100-101).*

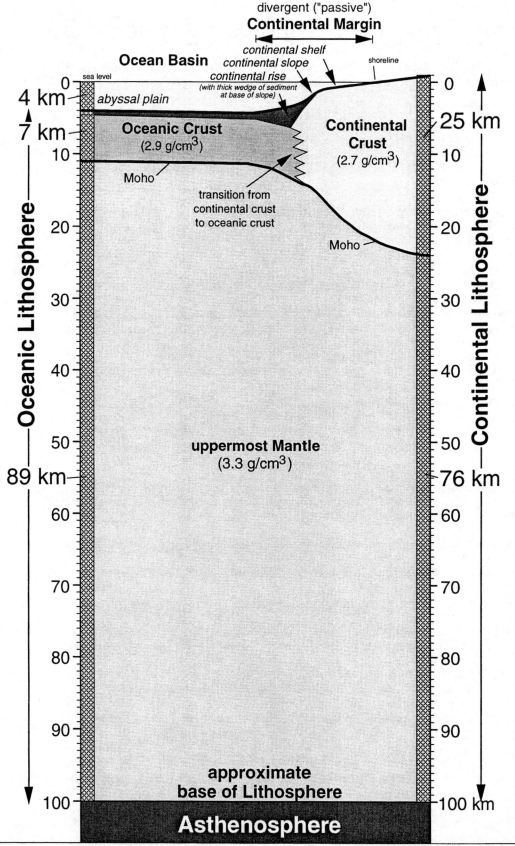

divergent ("passive")
Continental Margin

Ocean Basin

continental shelf
continental slope
continental rise
(with thick wedge of sediment at base of slope)

shoreline

sea level

4 km

abyssal plain

Oceanic Crust
(2.9 g/cm^3)

7 km

Continental Crust
(2.7 g/cm^3)

Moho

transition from
continental crust
to oceanic crust

Moho

Oceanic Lithosphere

Continental Lithosphere

uppermost Mantle
(3.3 g/cm^3)

89 km

76 km

**approximate
base of Lithosphere**

100 km

Asthenosphere

<u>**isostatic equilibrium**</u> maintained between the column of **Continental Lithosphere**
and the column of **Oceanic Lithosphere** by the underlying **Asthenosphere**

Continental Margins & Ocean Basin Physiography

The **physiography** (physical features) of the seafloor can be divided into two major provinces: Continental Margins and Ocean Basins.

Continental Margins

Continental margins represent the **transition from continental crust to oceanic crust**; the "step" from thick, buoyant continental crust to thinner, denser oceanic crust. **"Passive" continental margins** are not associated with plate boundaries and consist of three distinctive parts (shelf, slope, and rise). **"Active" continental margins** lack the continental rise because a deep-sea trench occurs at the base of the slope.

1. **continental shelf**
   - offshore extension of the continent (underlain by continental crust)
   - relatively shallow, gentle seaward slope (typically <1°)
   - shelf-slope break: 120-200 m water depth
2. **continental slope**
   - variable slope (gentle to steep escarpments)
   - underlain by thinned continental crust
   - dissected by numerous **submarine canyons** (deep erosional gullies, conduits for downslope sediment transport to the deep-sea)
3. **continental rise** *(only found along "passive" continental margins)*
   - thick accumulation of terrigenous sediment at the base of the slope, derived from the erosion of the continents (= **deep-sea fans**)
   - underlain by the boundary between continental and oceanic crust
4. **accretionary prism** *(only found along "active" continental margins)*
   - zone at trench where sediments are squeezed due to plate convergence (some sediment scraped-off subducting plate)
   - high fluid flow and chemical recycling (affects chemical mass balance of the ocean)
5. **fore-arc basin** *(only found along "active" continental margins)*
   - thick accumulation of terrigenous sediment between trench and magmatic arc

Ocean Basins

The ocean basins are underlain by **oceanic crust** composed of basalt. The ocean basins have many distinctive physiographic features related to plate tectonics (volcanism, seafloor spreading, subduction), hotspot volcanism, and reef development in the tropics.

1. **abyssal plains**
   - extensive in area, virtually flat (except where interrupted by volcanoes)
   - ~4000-6000 m water depth
2. **volcanoes** (various sizes)
   - **volcanic islands** (above sea level = recently active*)
   - **atolls** (steep-sided ring of coral reefs enclosing a lagoon; volcanic islands eventually drown* due to thermal subsidence)
   - **guyots** (flat-topped seamounts, formerly atolls at sea level*)
   - **seamounts** (submarine volcanic peaks)
   - **abyssal hills** (small volcanic peaks)
   *volcanoes will slowly subside (contract and sink) once their magma source is gone
3. **trenches** *(subduction zones at **convergent plate boundaries**)*
   - long, narrow, steep-sided troughs (e.g., Peru-Chile Trench)
   - deepest parts of ocean (>6000 m; Mariana Trench = 11,022 m)
   - seismically active (shallow to very deep earthquakes landward of trench)
   - associated with volcanism landward of trench: either **island arcs** or **volcanic mountain ranges** on land
4. **oceanic ridges & rises** *(spreading centers at **divergent plate boundaries**)*
   - active volcanic mountain ranges rising 2-3 km above the abyssal plains (e.g., Mid-Atlantic Ridge, East Pacific Rise)
   - continuous through ocean basins (>65,000 km around Earth)
   - extensive hydrothermal activity and chemical recycling of seawater through oceanic crust (important component of chemical mass balance of the ocean)
   - fault scars cut across and off-set these features (= **transform faults** and **fracture zones**)
   - seismically active (shallow earthquakes)

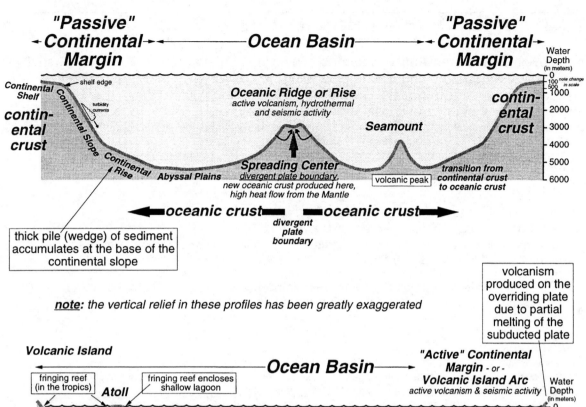

"Passive" Continental Margin ↔ **Ocean Basin** ↔ **"Passive" Continental Margin**

Water Depth (in meters)

Continental Shelf — shelf edge

continental crust

Continental Slope
Continental Rise
turbidity currents

Abyssal Plains

Oceanic Ridge or Rise
active volcanism, hydrothermal and seismic activity

Seamount

continental crust

Spreading Center
divergent plate boundary, new oceanic crust produced here, high heat flow from the Mantle

volcanic peak

transition from continental crust to oceanic crust

0
100 *note change in scale*
500
1000
2000
3000
4000
5000
6000

◄—oceanic crust—■   ■—oceanic crust—►

divergent plate boundary

thick pile (wedge) of sediment accumulates at the base of the continental slope

note: the vertical relief in these profiles has been greatly exaggerated

volcanism produced on the overriding plate due to partial melting of the subducted plate

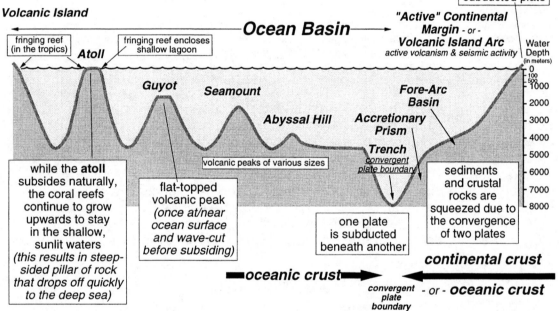

Volcanic Island

Ocean Basin

"Active" Continental Margin - or - **Volcanic Island Arc**
active volcanism & seismic activity

Water Depth (in meters)

fringing reef (in the tropics)

Atoll

fringing reef encloses shallow lagoon

Guyot   **Seamount**

Abyssal Hill

Fore-Arc Basin

Accretionary Prism

Trench
convergent plate boundary

0
100
500
1000
2000
3000
4000
5000
6000
7000
8000

volcanic peaks of various sizes

while the **atoll** subsides naturally, the coral reefs continue to grow upwards to stay in the shallow, sunlit waters *(this results in steep-sided pillar of rock that drops off quickly to the deep sea)*

flat-topped volcanic peak *(once at/near ocean surface and wave-cut before subsiding)*

one plate is subducted beneath another

sediments and crustal rocks are squeezed due to the convergence of two plates

continental crust

■—oceanic crust—►   ◄—

convergent plate boundary - or - **oceanic crust**

	Plate Tectonic Regime	Plate Boundary?	Volcanism?	Seismic Activity (Earthquakes)?
Passive Continental Margins (see p. 103)	Divergent	No	Generally None	Infrequent
Active Continental Margins (see p. 103)	Convergent	Yes, Deep-Sea Trench	Yes, Volcanic Mountain Chain (see p. 106-107)	Fairly Frequent; Shallow, Intermediate, & Deep-Focus

Plate Tectonics II: Plate Boundaries & Volcanism

High heat flow in the **Asthenosphere** causes convective motion (see p. 96-97). Material rises towards the surface in some places (upwelling) and sinks back into the mantle in others (downwelling). **Lithospheric Plates** are rafted by the ductile flow in the Asthenosphere. **Movement** of Lithospheric Plates is due to:

1. **gravitational sliding** off spreading center (ridge) as new oceanic crust cools and contracts (i.e., the oceanic crust becomes denser as it moves away from its volcanic source),
2. **slab pull** of dense subducting plates, and
3. **convective motion** in Asthenosphere (the rocks of the Asthenosphere flow laterally and drag the rigid Lithosphere along for the ride)

Recognition of lithospheric **plate boundaries**:

1. **earthquakes** (occur at *all 3 types of boundaries*):
    a. convergent plate boundaries (**deep-sea trenches**; shallow-, intermediate-, and deep-focus earthquakes along the **subduction zone**):
        • ocean-ocean collision (**island arcs**)
        • ocean-continent collision (**volcanic mountain chain**)
        • continent-continent collision (**massive mountain ranges**)
    b. divergent plate boundaries (**spreading centers**; shallow-focus earthquakes)
    c. strike-slip boundaries (**transform faults**; shallow-focus earthquakes)
2. **volcanism**
    a. at **spreading centers**, associated with rising (upwelling) mantle plumes
    b. landward of trenches: partial melting of subducting plate (downwelling) = magma production and volcanism on the overriding plate (formation of **magmatic arcs**):
        • **volcanic island arcs** *(islands and seamounts formed by ocean-ocean collisions)*
        • **volcanic mountain chains** *(formed by ocean-continent collisions)*

"Hot spots" are another type of volcanism unrelated to convection in the Asthenosphere and plate tectonics. Most hot spots represent **intraplate volcanism** (within a plate) which produce a **linear chain of volcanic islands or seamounts**. These chains show a progression in age from older to younger volcanoes as a lithospheric plate passes over a fixed (stationary) hot spot whose magma source lies deep in the Mantle. For example, the Hawaiian Islands, located far from any deep-sea trenches in the central North Pacific, were formed by a hot spot. The Emperor Seamount Chain-Hawaiian Ridge records the movement of the Pacific Plate over the hot spot during the past 80 million years. Hot spot tracks **do not delineate lithospheric plate boundaries**. See below.

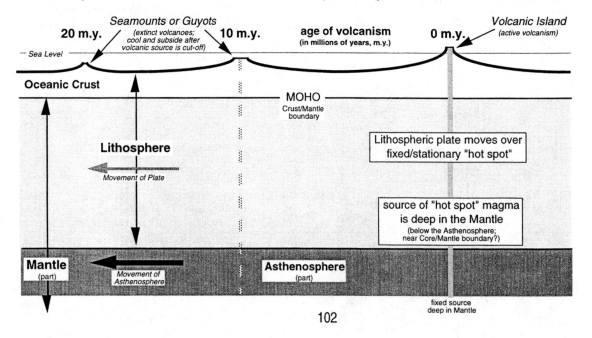

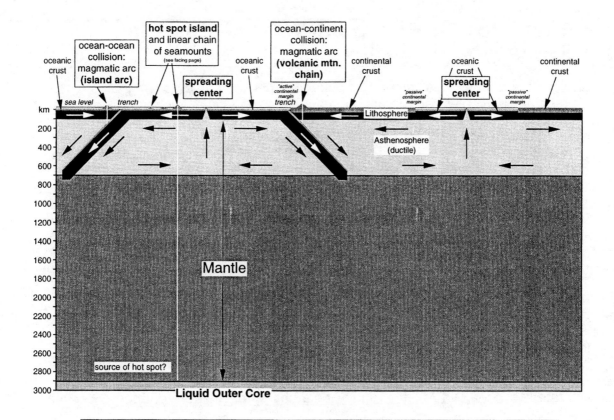

The following labels appear on the diagram:

oceanic crust

ocean-ocean collision: magmatic arc (island arc)

hot spot island and linear chain of seamounts (see facing page)

spreading center

oceanic crust

ocean-continent collision: magmatic arc (volcanic mtn. chain)

continental crust

oceanic crust

spreading center

continental crust

sea level   *trench*   *"active" continental margin trench*   *"passive" continental margin*   *"passive" continental margin*

km   200   400   600   800   1000   1200   1400   1600   1800   2000   2200   2400   2600   2800   3000

Lithosphere

Asthenosphere (ductile)

Mantle

source of hot spot?

Liquid Outer Core

Volcanism
major types and sources of magma

1. magmatic arcs *(convergent plate boundaries)*
- **volcanic mountain chains** (ocean-continent collisions) and **island arcs** (ocean-ocean collisions)
- <u>subduction</u> of oceanic lithosphere results in: 1) partial melting of subducting plate, and 2) volcanism on over-riding plate due rising of hot, buoyant magma
- <u>partial melting</u> of subducted crust + sediments (presence of water lowers the melting point of the rocks)
- **near plate margin**

2. spreading centers *(divergent plate boundaries)*
- "mid-ocean ridges"
- <u>convection</u> in Asthenosphere creates upwelling of mantle; adiabatic decompression causes the rocks to melt; rising of hot, buoyant magma produces volcanism at the seafloor
- **at plate margin**

3. "hot spots" *(not related to plate boundaries)*
- **linear chains** of islands, seamounts, or ridges
- plates move over <u>stationary</u> hot spot rooted deep in the mantle; adiabatic decompression causes the rocks to melt; rising of hot, buoyant magma produces volcanism at the seafloor
- plates continue to move over stationary hot spot resulting in linear chains of extinct, dormant, and active volcanoes
- most are **intraplate** (within the plate) rather than plate margin

Plate Tectonics III: Divergent & Strike-Slip Plate Boundaries

Spreading centers are the sites where new oceanic crust is created as two plates move apart. The spreading centers are a continuous chain of undersea volcanic mountains that extends from the Arctic Ocean, through the Atlantic, Indian, and Pacific Oceans for some 65,000 km. The **Mid-Atlantic Ridge** and the **East Pacific Rise** are parts of this continuous active volcanic mountain range on the seafloor. The spreading centers are situated over sites of **upwelling** in the Asthenosphere. Pressure is reduced as mantle rocks flow upward towards the surface. This release of pressure, called **adiabatic decompression**, causes the rocks to partially melt. Melted rock is called **magma**. The hot, buoyant magma rises and collects near the surface in **magma chambers**. Some of the magma cools and crystallizes within the crust to form a coarse-grained rock called **gabbro**. Some of the magma is extruded at the surface as **lava**, which rapidly cools when it comes in contact with the cold seawater to form a fine-grained rock called **basalt**.

High heat flow from the Mantle causes the spreading centers to stand high above the adjacent abyssal plains. The shape of the ridge profile is related to the **rate of seafloor spreading**. Parts of the global ridge system with high rates of upwelling from the Asthenosphere produce relatively large magma chambers while other parts are characterized by much smaller magma chambers. The East Pacific Rise is creating new ocean crust at an average rate of about 6 cm/year for each flank of the ridge (=half-spreading rate) while the Mid-Atlantic Ridge has a half-spreading rate of about 3 cm/year.

As an old continent is split (**rifted**) into two, a new ocean basin, composed of ocean crust (basalt), is created in the growing void between the rifted continental fragments. **Passive continental margins** form at the transition between the thinned, stretched continental crust and the new oceanic crust. The new ocean basin has an active spreading center (divergent plate boundary) approximately equidistant between the two rifted continental margins. The Red Sea is an example of a young and growing ocean basin.

Because the Lithosphere is rigid, stresses on the crustal rocks that cover our spherical planet cause the crust to crack. There are numerous cracks along the spreading centers because not all segments of the volcanic ridge system are active at the same time imparting differential strain on the brittle oceanic crust. In addition to these stresses, subduction of oceanic crust is associated with so much friction that the process is far from continuous or smooth. Strain builds up where two plates collide; occasionally the strain is released catastrophically and seismic energy is released as a crustal segment lurches forward. A consequence of these differential stresses on the oceanic crust is the multitude of faults that offset the spreading centers. Faults mark places where rocks are cracked and where there is movement along the cracks. **Transform faults** are the cracks in the ocean crust that are responsible for the offsets observed along the spreading centers. Notice that where the ridge is offset by a transform fault, the sense of motion is opposite on either side of the fault. Transform faults have lateral, or **strike-slip motion**, where two lithospheric plates move past one another making these regions **prone to seismic activity**. This opposite sense of motion accumulates strain until the rock can no longer accommodate the pressure and the frictional forces snap catastrophically to produce an earthquake. The seismically active **San Andreas Fault System** of southern California is a transform fault between the **East Pacific Rise** (Baja California) and the **Juan de Fuca Ridge** off Oregon and Washington. **Fracture zones** are long, linear zones of highly fractured oceanic crust. They represent continuations of transform faults beyond the zone of active strike-slip displacement. Fracture zones tend to be **seismically inactive** because the sense of motion on both sides of the fracture is in the same direction.

Divergent Plate Boundaries

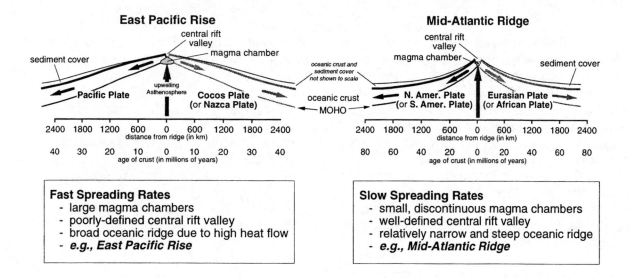

East Pacific Rise

- sediment cover
- central rift valley
- magma chamber
- Pacific Plate
- upwelling Asthenosphere
- Cocos Plate (or Nazca Plate)

oceanic crust and sediment cover not shown to scale

oceanic crust
— MOHO —

Mid-Atlantic Ridge

- central rift valley
- magma chamber
- sediment cover
- N. Amer. Plate (or S. Amer. Plate)
- Eurasian Plate (or African Plate)

| 2400 | 1800 | 1200 | 600 | 0 | 600 | 1200 | 1800 | 2400 |

distance from ridge (in km)

| 40 | 30 | 20 | 10 | 0 | 10 | 20 | 30 | 40 |

age of crust (in millions of years)

| 2400 | 1800 | 1200 | 600 | 0 | 600 | 1200 | 1800 | 2400 |

distance from ridge (in km)

| 80 | 60 | 40 | 20 | 0 | 20 | 40 | 60 | 80 |

age of crust (in millions of years)

Fast Spreading Rates
- large magma chambers
- poorly-defined central rift valley
- broad oceanic ridge due to high heat flow
- *e.g., East Pacific Rise*

Slow Spreading Rates
- small, discontinuous magma chambers
- well-defined central rift valley
- relatively narrow and steep oceanic ridge
- *e.g., Mid-Atlantic Ridge*

Strike-Slip Plate Boundaries

*Oblique Map View of a Oceanic Ridge System (= **Spreading Center**):*
*Note the two ridge offsets (= **Transform Faults**), also note that the*
two plates are moving in opposite directions on either side of the fault.

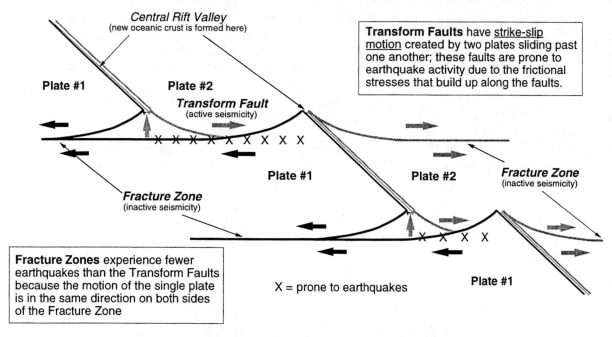

Central Rift Valley
(new oceanic crust is formed here)

Transform Faults have strike-slip motion created by two plates sliding past one another; these faults are prone to earthquake activity due to the frictional stresses that build up along the faults.

Plate #1

Plate #2

Transform Fault
(active seismicity)

Plate #1

Plate #2

Fracture Zone
(inactive seismicity)

Fracture Zone
(inactive seismicity)

Fracture Zones experience fewer earthquakes than the Transform Faults because the motion of the single plate is in the same direction on both sides of the Fracture Zone

X = prone to earthquakes

Plate #1

*The **San Andreas Fault** system through southern California is a transform fault between two active spreading centers (the **East Pacific Rise** to the southeast and the **Juan de Fuca Ridge** to the northwest).*

Plate Tectonics IV: Convergent Plate Boundaries

Convergent plate boundaries are associated with **two diagnostic features**: 1) seismic activity, and 2) volcanism and/or mountain building (**orogeny**). The seismic activity is a consequence of subduction of oceanic lithosphere and/or faulting due to collisional tectonics. The **deep-sea trenches** mark plate boundaries where oceanic lithosphere is subducted beneath younger oceanic lithosphere (**ocean-ocean collision**), or beneath continental lithosphere (**ocean-continent collision**). A **continent-continent collision** often begins as an ocean-continent collision. Once the thick continental lithosphere collides with another block of continental lithosphere, subduction becomes minimal and massive mountains grow vertically as the two plates are pushed towards each other. Deep-sea sediments and slices of oceanic crust are squeezed and deformed in the **suture zone** which represents all that is left of the ocean basin that once existed between the two continental land masses. The **Himalaya Mountains** and **Tibetan Plateau**, the largest mountains in the world today, are still growing taller as the subcontinent of India drives northward into the continent of Asia. The old **Appalachian Mountains** is another example of a mountain range produced by continent-continent collision.

The volcanism characteristic of ocean-ocean and ocean-continent collisions is the consequence of subduction of oceanic lithosphere. The **partial melting** of oceanic crust and deep-sea sediments subducted at the trench is facilitated by the presence of water, which lowers the melting temperature of rocks. The resulting pools of melted rock are called **magma**. This magma is less dense than the parent rock, and since hot rock is less dense than cold rock, the magma rises buoyantly through the lithosphere of the overriding plate. Some of this magma cools within the crust to form solid rock such as **granite**, but some of the magma reaches the surface where it erupts as a volcano. The new accumulation of volcanic rock is called a **magmatic arc**. In the case of an ocean-ocean collision, the resulting volcanism produces an arcuate chain (arc-like) of volcanic islands and seamounts referred to as an **island arc**. The **Aleutian Islands** and the **Mariana Islands** of the North Pacific, and the **Lesser Antilles** of the Caribbean are all examples of island arcs. The volcanoes are not continuously active, but the volcanic chain is approximately the same age all along its length (this is in contrast to the linear chain of volcanoes produced by a stationary hot spot which displays a steady progression of age from older to younger volcanoes towards the single hot spot source). In the case of an ocean-continent collision, a **chain of volcanic mountains** is produced landward of the trench near the edge of the overriding continent. The **Andes Mountains** of South America and the **Cascade Mountains** of Washington and Oregon (including Mt. Rainer, Mt. St. Helens, Mt. Hood) are excellent examples of such volcanic mountain chains.

Convergent Plate Boundaries

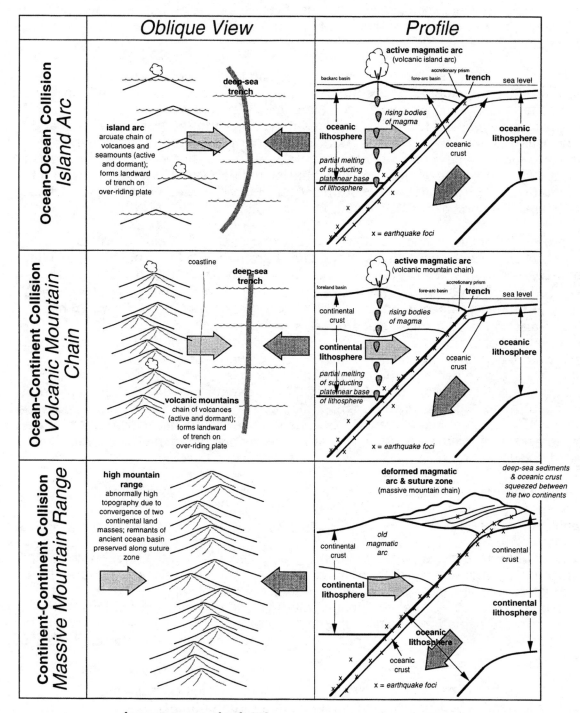

	Oblique View	*Profile*
Ocean-Ocean Collision *Island Arc*	**deep-sea trench** — **island arc** arcuate chain of volcanoes and seamounts (active and dormant); forms landward of trench on over-riding plate	**active magmatic arc** (volcanic island arc); backarc basin, fore-arc basin, accretionary prism, **trench**, sea level; **oceanic lithosphere**, *rising bodies of magma*, oceanic crust, **oceanic lithosphere**; *partial melting of subducting plate near base of lithosphere*; x = earthquake foci
Ocean-Continent Collision *Volcanic Mountain Chain*	coastline, **deep-sea trench** — **volcanic mountains** chain of volcanoes (active and dormant); forms landward of trench on over-riding plate	**active magmatic arc** (volcanic mountain chain); foreland basin, fore-arc basin, accretionary prism, **trench**, sea level; continental crust, **continental lithosphere**, *rising bodies of magma*, oceanic crust, **oceanic lithosphere**; *partial melting of subducting plate near base of lithosphere*; x = earthquake foci
Continent-Continent Collision *Massive Mountain Range*	**high mountain range** abnormally high topography due to convergence of two continental land masses; remnants of ancient ocean basin preserved along suture zone	**deformed magmatic arc & suture zone** (massive mountain chain); *deep-sea sediments & oceanic crust squeezed between the two continents*; continental crust, *old magmatic arc*, continental crust; **continental lithosphere**, **oceanic lithosphere**, oceanic crust, **continental lithosphere**; x = earthquake foci

large arrows depict the convergence of two lithospheric plates

107

Marine Sediments

Sediments on the floor of the deep-sea are classified according to the **source** or **composition** of the materials. Their **distribution** in the world ocean is related to a number of factors including proximity to source, processes of distribution (gravity, deep and shallow ocean currents, and wind), and ocean chemistry. Mixtures of sediment types are common.

Terrigenous or **lithic sediment** is composed of sand, silt, or clay-sized particles derived from the **physical and chemical weathering** of rocks and soil on land. These sediments form an apron of debris around the continents, consisting mostly of sand and mud. Specific *varieties* of terrigenous sediment include **red clay** (wind-blown silt and clay deposited on the abyssal plains), **neritic sediment** (referring to terrigenous sediment found on continental shelves), **glacial marine sediment** (sediment deposited by glacial ice or transported out to sea by icebergs and deposited in an apron around the high latitudes), and **volcaniclastic sediment** (eroded or ejected volcanic debris and ash deposited around volcanic islands and seamounts).

Biogenic sediment is composed of the microscopic shells of marine **plankton**, typically protists. Plankton with mineralized shells are particularly important contributors to deep-sea sediments. Plankton with shells of **calcium carbonate** ($CaCO_3$) produce a type of sediment called **calcareous ooze**, and plankton with shells of **opaline silica** ($SiO_2.H_2O$) produce **siliceous ooze**. Plankton are grazed and preyed upon by many types of small and large animals. As a result, their microscopic shells are packaged into **fecal pellets**. Fecal pellets are an important mode of transport of the tiny shells from the surface waters where the plankton live to the seafloor where their empty shells accumulate. Passive settling through the water column is another mode of deposition. The **Carbonate Compensation Depth** (**CCD**) represents a chemical boundary in the deep ocean (~4000-5000 m water depth). Calcareous ooze does not accumulate on the seafloor at depths greater than the CCD because of intense chemical dissolution caused by low temperature, high pressure, and relatively high concentration of dissolved CO_2 (see p. 154-155).

Authigenic sediment precipitates directly from seawater. Authigenic sediments are most common in areas below the CCD, or in areas of very slow pelagic or terrigenous accumulation rates, or beneath areas of high biological productivity (see below).

Pelagic sediment refers to those sedimentary particles which settle through the water column, such as shells of plankton and wind-blown silt and clay (biogenic ooze and red clay are pelagic sediments). **Hemipelagic sediment** is typically a mix of terrigenous and biogenic sediments such as terrigenous mud + calcareous ooze, or red clay + siliceous ooze. **Periplatform ooze** is the carbonate sediment, rich in shallow water biogenic shell debris and calcareous ooze that accumulates in an apron around reefs and atolls.

The rate at which sediments accumulate in the deep-sea varies considerably. For example, terrigenous sediment accumulates at rates (greater than) >5 cm/kyr (centimeters per thousand years) along many parts of the continental margin. Biogenic ooze typically accumulates at rates of 1-3 cm/kyr, although higher rates can occur in areas of very high biological productivity. Abyssal red clay accumulates at rates (much less than) <<1 cm/kyr, or about the rate at which dust accumulates in your home.

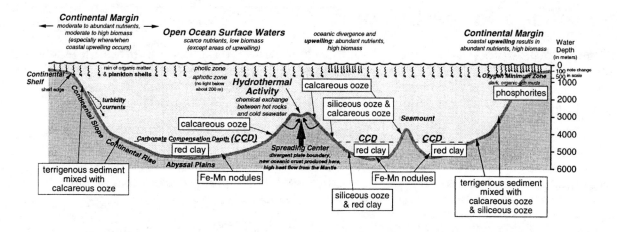

The following represent the major sediment types found on the floor of the deep-sea (*mixtures of sediment types are common*):

Terrigenous Sediment

- derived from **weathering of rocks** on land
- sand and mud are moved across **continental shelf** by large storms and gravity
- much sediment is funneled into **submarine canyons** at the edge of the continental shelf
- transport down the **continental slope** via slumps and **turbidity currents**
- terrigenous sediment builds-up as **deep-sea fans** at the base of the continental slope
- the depositional feature at the base of the slope is called the **continental rise**

Terrigenous sediment accumulates along the continental margins; terrigenous sediment typically masks other sediment types close to the continents because of high sedimentation rates.

Red Clay

- red or brown clay is derived from **wind-blown dust** which slowly settles to the seafloor, or it is derived from deep current-transported clays
- red clay **accumulates at very slow rates** and is diluted by other types of sediment in many areas of the deep-sea

Red clay accumulates on the abyssal plains, deeper than 4500 meters (below CCD).

Biogenic Sediment
1. Calcareous Ooze

- composed of **carbonate ($CaCO_3$) shells of plankton**
- dead plankton settle through the water column, or consumed plankton are incorporated into **fecal pellets** which settle to the seafloor
- calcium carbonate dissolves at about 4500 m water depth due to cold temperature, high pressure, and greater CO_2 content of deep waters = **carbonate compensation depth (CCD)**

Calcareous ooze accumulates on bathymetric highs beyond continental margins, shallower than 4500 meters (above CCD), such as the spreading centers and flanks of volcanoes that stick up above the abyssal plains.

2. Siliceous Ooze

- composed of **siliceous (SiO_2) shells of plankton**
- **areas of upwelling** provide the nutrients and silica necessary to sustain a large biomass of siliceous plankton

Siliceous ooze accumulates beneath areas of high biological productivity in the surface ocean, such as the equatorial Pacific and around Antarctica

Authigenic Sediment

- **ferromanganese nodules** form on the abyssal plains by the slow chemical precipitation of metal oxides directly from seawater (facilitated by biochemical activity of bacteria and other microorganisms, and by the burrowing activity of larger organisms)
- **hydrothermal activity** at spreading centers is probably a major source of dissolved metal oxides
- **phosphorite nodules** form on the outer shelf and upper slope where unoxidized organic matter in the **oxygen minimum zone** is biochemically transformed into phosphorite
- some authigenic deposits are **economically valuable**

Authigenic sediment precipitates directly from seawater under specific (bio)chemical conditions related to source, oxidation-reduction, and/or sedimentation rate.

The Unique Properties of Water

The **water molecule (H₂O)** has a **"dipolar" structure**. This means that the molecule has an uneven distribution of electrical charge owing to the asymmetrical shape of the water molecule (it looks like a "mickey mouse" head). Instead of the **two Hydrogen atoms** being bonded on either side of the **single Oxygen atom** (i.e., 180° apart), the Hydrogen atoms are 105° apart. In this configuration the Hydrogen-side of the molecule (**H₂**) has a net **positive charge** because more of the positively charged protons are concentrated on this side of the molecule. The Oxygen-side (**O**) has a net **negative charge** because of the greater concentration of negatively charged electrons around the O atom.

Because of this asymmetrical distribution of electrical charge in the water molecule, adjacent water molecules may be attracted to one another. **Hydrogen bonds** are electrostatic forces between water molecules (see facing page). The proportion of water molecules bonded to other water molecules is related to the gain and loss of heat, which is in the form of **sensible heat** and **latent heat** (discussed below). The dipolar structure of the water molecule and the formation of hydrogen bonds between water molecules are responsible for the many unique properties of water:

1. high surface tension (cohesion)
   - H bonds allow water to form droplets (e.g., raindrops)

2. water occurs in **3 states (phases)**
   - water is the only naturally occurring substance at Earth's surface and atmosphere to exist simultaneously in all 3 phases: **solid, liquid,** and **gas**

3. water is a **good solvent**
   - dipolar structure weakens attraction between other molecules (e.g., NaCl)

4. "high" boiling and freezing points
   - compared with other H₂X molecules lacking the dipolar structure (e.g., H₂S, H₂Se, and H₂Te)
   - in water, H bonds must be broken or formed (occurs at much higher temperatures than other H₂X molecules)

5. water has high **heat capacity**
   - compared with many other substances (liquid and solid), water absorbs and releases a considerable amount of heat with little change in temperature
   note: *heat is measured in* **calories** *(calorie = amount of heat needed to raise the temperature of 1g of water by 1°C);* **sensible heat** *= vibration of water molecules, the temperature measured with a thermometer*

> * water's high heat capacity helps to moderate surface temperatures on Earth

6. high **latent heat of vaporization** and high **latent heat of fusion**
   - considerable amounts of heat must be added or removed to convert water from one phase to another (**ice** ↔ **water** ↔ **vapor**)
   note: *latent heat = the heat required to change phase* without a change of temperature
   - 540 calories/g are required to **evaporate water** and 80 calories/g are required to **melt ice**
   - 540 calories/g of latent heat are released to the atmosphere when **water vapor condenses** to form clouds, rain, and snow, and 80 calories/g are released when **water freezes**

> * the latent heat required to break hydrogen bonds (melt ice or evaporate water), and the latent heat released to the atmosphere when H-bonds reform (condense water vapor to form liquid water droplets or ice) are important processes in the hydrologic cycle, which moves moisture and heat around the planet (see p. 112-113)

Water Molecule (H₂O)

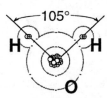

2 Hydrogen (H) atoms, each with 1 <u>proton</u> (shaded) and 1 <u>neutron</u> (clear) in the nucleus, and 1 <u>electron</u> (-) in orbit around the nucleus

Oxygen (O) atom with 8 protons and 8 neutrons in the nucleus, and 8 electrons in orbit around the nucleus (2 electrons in the inner orbit and 6 in the outer orbit)

Water molecule held together by strong **covalent bonds:** each **H** shares its one (-) with the **O** to completely fill the outer electron shell of the **O** with 8 electrons, and the **O** shares (-) to fill the single electron shell of each **H** with 2 (-).

Dipolar Structure

Asymmetrical distribution of electrical charge results in the H-end being more positive and the O-end more negative.

Hydrogen Bonds

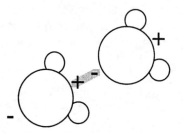

Hydrogen bonds are the relatively weak electrostatic forces that cause attraction between oppositely charged ends of adjacent water molecules.

Water occurs in 3 different phases

Solid (ice)
(3-D crystalline structure)

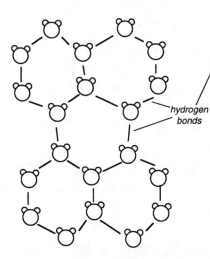

crystalline structure is less dense than liquid water

Liquid (water)

surface tension at the air/water interface

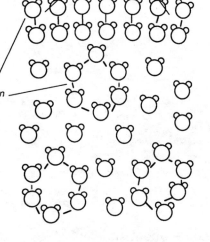

hydrogen bonds

liquid water is more dense than ice

Gas (water vapor)

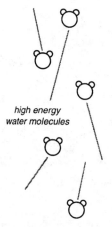

high energy water molecules

high energy molecules with few H-bonds

Hydrologic Cycle & Latent Heat: the Movement of Water & Heat around the Planet

As water molecules absorb energy from the Sun or other source, some of the energy causes the water molecules to vibrate faster. This is called **sensible heat** and it represents the temperature measured with a thermometer. Much of the remaining heat causes **hydrogen bonds (H bonds)** to be broken resulting in a *change of phase without a change in temperature* (e.g., the gradual melting of ice cubes in your soda or the evaporation of water from a puddle on the road whether the day is hot or cold). This is called **latent heat**. The opposite is true when water molecules are cooled: they vibrate less vigorously as they lose sensible heat, and they release latent heat to the atmosphere as H bonds are reformed (e.g., water vapor condenses to form water droplets and clouds, or liquid water begins to form ice crystals).

Considerable heat energy is required to **break hydrogen bonds** when <u>ice melts</u> (80 calories per gram of water, cal/g) or when <u>water evaporates</u> (540 cal/g). This "extra" heat is called the **latent heat of vaporization**. In the opposite way, heat energy is released to the atmosphere as **hydrogen bonds form** when <u>water freezes</u> (80 cal/g) or when <u>water vapor condenses</u> (540 cal/g) to form clouds or precipitation. This is called the **latent heat of fusion**.

The **hydrologic cycle** is a major player in the redistribution of heat around the planet by way of cycling water between the ocean, atmosphere, and land. **Evaporation** is the process whereby latent heat is removed from the ocean and land, while **condensation and precipitation** release latent heat to the atmosphere. **Transpiration** is an important process by which plants release water back to the atmosphere through their leaves. **Runoff** includes both surface (streams and rivers) and subsurface (groundwater) flow of liquid water toward the ocean under the influence of gravity or hydrostatic pressure. These processes of the hydrologic cycle help to redistribute excess tropical heat to colder regions by way of the atmosphere and day-to-day weather (see p. 136). **Tropical cyclones (hurricanes and typhoons)** represent safety valves for the release of excess heat that builds up in the tropics and subtropics every year.

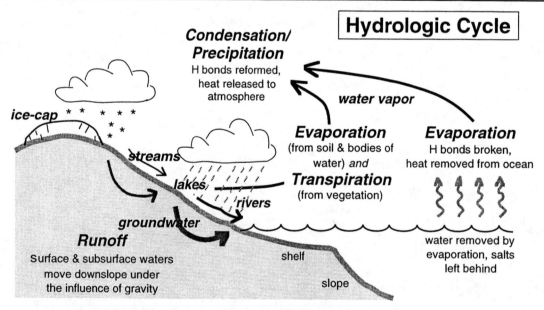

The atmosphere and ocean share about equally in the redistribution of heat around the planet through the processes of the hydrologic cycle (evaporation and precipitation) and surface currents (see p. 116-117).

Sensible & Latent Heat

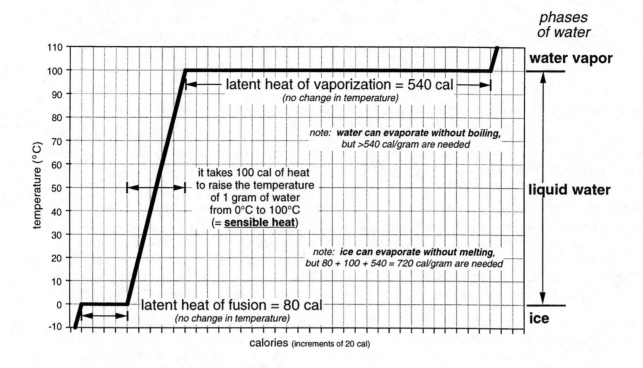

sensible heat versus latent heat of vaporization and latent heat of fusion

- **latent heat** is the "*extra*" *heat* added (or removed) to <u>change water from one phase to another</u>; this occurs **without a change in temperature** because the energy gained (or lost) is used to break (or form) hydrogen bonds rather than increase (or decrease) the <u>rate of vibration of the water molecules</u> which is **sensible heat**

- this "additional heat", or **latent heat**, is necessary to **break H bonds** during evaporation of water, melting of ice, & sublimation of ice, or the latent heat released when **H bonds form** with condensation/precipitation or freezing

- **540 calories** (per gram of water) of latent heat are *needed to evaporate water* and **540 calories** of latent heat are *released to the atmosphere when water vapor condenses* to form clouds, rain, and snow

- **80 calories** (per gram of water) of latent heat are *needed to melt ice* and **80 calories** are *released to the atmosphere when water freezes*

- latent heat is important for the **hydrologic cycle**, which moves moisture and heat around the planet

Seawater Salinity: the Salt of the Ocean

There are many substances dissolved in seawater including 1) **gases** such as O_2 and CO_2, 2) **nutrients** required by organisms such as nitrates, phosphates, trace elements and vitamins, 3) dissolved **organic molecules**, and 4) **mineral salts**. **Salinity** refers to the <u>total dissolved solids</u> in water. Average **seawater** is 96.5% water and 3.5% total dissolved inorganic solids (**freshwater**, is essentially 100% water with only traces of dissolved substances). Oceanographers describe salinity with units of **parts per thousand** ($^o/_{oo}$; i.e., grams of dissolved solids per kilogram of water, g/kg) rather than parts per hundred which is percent (%). **Average ocean salinity is ~35$^o/_{oo}$** (fresh water is ~0$^o/_{oo}$). In most parts of the ocean, the salinity ranges between 33$^o/_{oo}$ and 37$^o/_{oo}$.

Many of the inorganic substances dissolved in seawater occur in **ionic form**. The dipolar nature of the water molecule has dissociated the bonds between these substances leaving them with either a positive charge or a negative charge. Positively charged ions are called **cations** and negatively charged ions are called **anions**. The two most common dissolved solids in seawater are **chloride (Cl^-)**, an anion that accounts for 55.1% by weight of the ocean's salinity, and **sodium (Na^+)**, a cation that accounts for 30.6%. These two ions combine when seawater is evaporated to form **NaCl** (sodium chloride) which is <u>common table salt</u>. The other common ions in seawater are **sulfate (SO_4^{2-}, 7.7%), magnesium (Mg^{2+}, 3.7%), calcium (Ca^{2+}, 1.2%)**, and **potassium (K^+, 1.1%)**. These 6 ions account for 99.4% of all the dissolved inorganic solids in seawater.

A major source of the dissolved solids in seawater is from the chemical weathering of rocks and soils on land. The ions are transported to the ocean by rivers (**runoff**). If you compare the proportion of dissolved ions in seawater with the proportion found in rivers (below), you'll notice that there is a significant difference. For example, the four most common ions in seawater (in descending order of abundance) are Cl^-, Na^+, SO_4^{2-}, and Mg^{2+}, while in rivers the top four dissolved ions are HCO_3^-, Ca^{2+}, SiO_2, and SO_4^{2-}. There are several reasons for this: 1) there are additional sources of dissolved solids in seawater besides runoff from the land, including **hydrothermal circulation** at the spreading centers (see facing page) and **volcanic emissions** (see p.), 2) many of the elements and substances dissolved in seawater are utilized by organisms in complex **biochemical cycles**, and 3) different ionic species have different **residence times** in the ocean. Residence time refers to the length of time an ion remains in seawater before it is used by organisms, buried in sediments, or involved in **geochemical cycles** at subduction zones (accretionary prisms) and spreading centers (hydrothermal vents). Ions that are involved with biochemical cycles such as Ca^{2+}, HCO_3^-, SiO_2, and **Fe** have relatively short residence times (200 years to 2 million years) while the most common ions in seawater, Cl^- and Na^+, are largely unreactive and therefore have very long residence times (hundreds of millions of years). The chemical composition of the ocean has remained relatively constant over time because input = output.

DISSOLVED ION	SEAWATER	RUNOFF (RIVERS)
chloride (Cl^-)	19.3$^o/_{oo}$ ($^o/_{oo}$ = g/kg)	0.008$^o/_{oo}$ (= 7.8 ppm)
sodium (Na^+)	10.8$^o/_{oo}$	0.006$^o/_{oo}$ (6.3 ppm)
sulfate (SO_4^{2-})	2.7$^o/_{oo}$	0.011$^o/_{oo}$ (11.2 ppm)
magnesium (Mg^{2+})	1.3$^o/_{oo}$	0.004$^o/_{oo}$ (4.1 ppm)
calcium (Ca^{2+})	0.4$^o/_{oo}$	0.015$^o/_{oo}$ (15.0 ppm)
potassium (K^+)	0.4$^o/_{oo}$	0.002$^o/_{oo}$ (2.3 ppm)
bicarbonate (HCO_3^-)	0.1$^o/_{oo}$	0.058$^o/_{oo}$ (58.4 ppm)
silicate (SiO_2)	0.003$^o/_{oo}$	0.013$^o/_{oo}$ (13.1 ppm)

1$^o/_{oo}$ (part per thousand) = 1 g salt/kg water (gram per thousand grams); 1$^o/_{oo}$ = 1000 ppm (part per million)

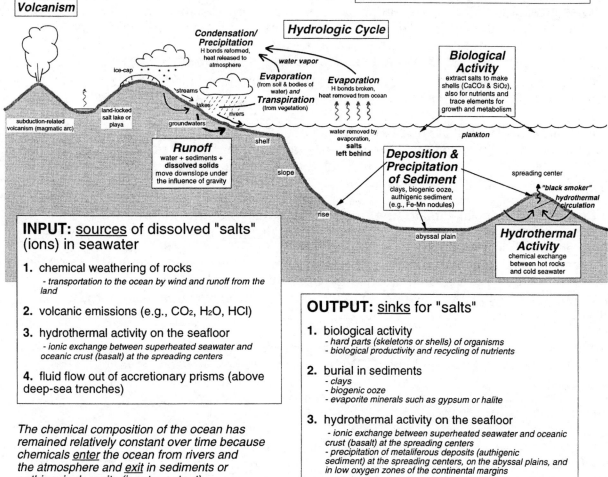

Sources and Sinks of Dissolved Solids in Seawater

Volcanism

Hydrologic Cycle

Condensation/ Precipitation
H bonds reformed, heat released to atmosphere

water vapor

ice-cap

Evaporation
(from soil & bodies of water) and
Transpiration
(from vegetation)

Evaporation
H bonds broken, heat removed from ocean

streams

lakes

rivers

groundwaters

subduction-related volcanism (magmatic arc)

land-locked salt lake or playa

Runoff
water + sediments + **dissolved solids** move downslope under the influence of gravity

shelf

slope

rise

abyssal plain

water removed by evaporation, **salts left behind**

Biological Activity
extract salts to make shells ($CaCO_3$ & SiO_2), also for nutrients and trace elements for growth and metabolism

plankton

Deposition & Precipitation of Sediment
clays, biogenic ooze, authigenic sediment (e.g., Fe-Mn nodules)

spreading center

"black smoker"
hydrothermal circulation

Hydrothermal Activity
chemical exchange between hot rocks and cold seawater

INPUT: sources of dissolved "salts" (ions) in seawater

1. chemical weathering of rocks
   - transportation to the ocean by wind and runoff from the land

2. volcanic emissions (e.g., CO_2, H_2O, HCl)

3. hydrothermal activity on the seafloor
   - ionic exchange between superheated seawater and oceanic crust (basalt) at the spreading centers

4. fluid flow out of accretionary prisms (above deep-sea trenches)

The chemical composition of the ocean has remained relatively constant over time because chemicals enter the ocean from rivers and the atmosphere and exit in sediments or authigenic deposits (input = output); enroute many elements and compounds are part of complex biological pathways

OUTPUT: sinks for "salts"

1. biological activity
   - hard parts (skeletons or shells) of organisms
   - biological productivity and recycling of nutrients

2. burial in sediments
   - clays
   - biogenic ooze
   - evaporite minerals such as gypsum or halite

3. hydrothermal activity on the seafloor
   - ionic exchange between superheated seawater and oceanic crust (basalt) at the spreading centers
   - precipitation of metaliferous deposits (authigenic sediment) at the spreading centers, on the abyssal plains, and in low oxygen zones of the continental margins

4. subduction
   - recycling of crust and sediments

Hydrothermal Circulation

In addition to rivers, dissolved solids enter seawater as it circulates through the fractured oceanic crust of the active spreading centers. Here cold, deep waters are drawn into the crust, superheated by the magma reservoir below the ridge, and then expelled as hot fluids. This **hydrothermal circulation** vents fluids of a different composition than the original seawater. For example, seawater Mg^{2+} is exchanged for Ca^{2+} from the basalt. Some vents expel hot fluids rich in dissolved metals (e.g., Fe, Mn, Cu, Zn, Co) extracted from interaction with the basalt. These **"black smokers"** occur in the central rift valley of active spreading centers. **Ionic exchange** between hydrothermal fluids and the oceanic crust is responsible for the metaliferous ions precipitated in authigenic deposits found on the seafloor, including **manganese nodules** (see p. 108-109).

Distribution of Heat: Inequity Drives Fluid Flow

The Sun is approximately 93 million miles (~150 million km) from Earth and its diameter is ~100 times greater than that of the Earth. Because of its great distance and size, solar energy reaches the Earth as parallel rays. However, the distribution of *incoming sol*ar radi*ation* (**insolation**) across Earth's surface is uneven because of the spherical shape of the planet. There are also significant seasonal changes in the amount of solar radiation received at any given latitude because Earth's axis of rotation is tilted 23.5° relative to its orbital plane around the Sun, called the **plane of the ecliptic** (see p. 119).

The **low latitudes** (tropical climate belt) receive abundant solar radiation throughout the year because of the <u>high angle of solar incidence</u> (sun is higher in the sky at mid-day). More solar radiation is received during daylight hours than is radiated back into space at night, so there is a **net gain of insolation**.

The **high latitudes** (polar climate belt) experience a **net loss of insolation** over the course of a year because of the <u>lower angle of solar incidence</u> (sun is lower in the sky at mid-day) and total darkness during the winter months (north of the Arctic Circle and south of the Antarctic Circle; see p. 118-119). The same amount of solar energy that strikes the low latitudes is dispersed across a much larger surface area in the high latitudes. Heat absorbed by the ocean and atmosphere in the high latitudes is less than the energy radiated back to space.

Gradients of temperature and pressure across the globe **drive fluid motions** that counteract the unequal heat distribution. Water masses and air masses <u>move from areas of high pressure to areas of low pressure</u>. The ocean and atmosphere share about equally in redistributing the excess heat of the tropics toward the heat-deficient polar regions.

1. The **atmosphere** transports heat by way of the **hydrologic cycle** (see p. 112-113). Hydrogen bonds are broken during the **evaporation** of water and <u>latent heat is removed from ocean</u>. As moisture-rich air masses cool, **condensation** of water vapor causes tiny water droplets to form, clouds to build, and precipitation to fall. During the process of condensation, hydrogen bonds reform between water molecules and <u>latent heat is released to the atmosphere</u>. In this way, tropical heat is redistributed to the higher latitudes.

2. The **ocean** transports heat by large **surface currents** of the upper water masses (see p. 130-131). For example, the **subtropical gyres** represent large circulation cells that transport warm waters poleward along the western sides of the ocean basins and cool waters equatorward along the eastern sides. Examples of this circulation pattern include the **Gulf Stream** (warm) and **Canary Current** (cool) in the North Atlantic, and the **Kuroshio Current** (warm) and **California Current** (cool) in the North Pacific.

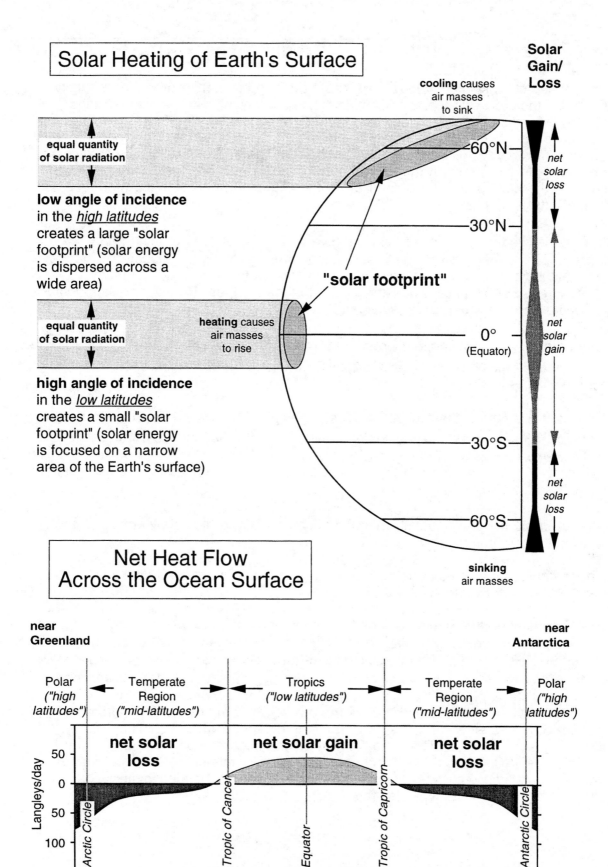

Solar Heating of Earth's Surface

Solar Gain/ Loss

cooling causes
air masses
to sink

equal quantity
of solar radiation

low angle of incidence
in the *high latitudes*
creates a large "solar
footprint" (solar energy
is dispersed across a
wide area)

—60°N—

—30°N—

*net
solar
loss*

"solar footprint"

equal quantity
of solar radiation

heating causes
air masses
to rise

0°
(Equator)

*net
solar
gain*

high angle of incidence
in the *low latitudes*
creates a small "solar
footprint" (solar energy
is focused on a narrow
area of the Earth's surface)

—30°S—

*net
solar
loss*

—60°S—

sinking
air masses

Net Heat Flow Across the Ocean Surface

near
Greenland

near
Antarctica

| Polar ("high latitudes") | ← Temperate Region ("mid-latitudes") → | ← Tropics ("low latitudes") → | ← Temperate Region ("mid-latitudes") → | Polar ("high latitudes") |

net solar loss | **net solar gain** | **net solar loss**

Langleys/day

50
0
50
100

Arctic Circle

Tropic of Cancer

Equator

Tropic of Capricorn

Antarctic Circle

70°N 60°N 50°N 40°N 30°N 20°N 10°N 0° 10°S 20°S 30°S 40°S 50°S 60°S 70°S

The Seasons: Behind the Wheel of the Climate Engine

The annual progression of the **seasons** has a profound influence on the distribution of heat during the course of a year. For example, some localities in the high latitudes experience weeks of total darkness, and then six months later experience weeks without night, while many localities in the mid-latitudes are characterized by bitterly cold winters and hot summers (see p. 128-129). These observations illustrate the yearly extremes of **heat gain and heat loss** at the Earth's surface. **Seasonality** also has a profound impact on the Earth's biosphere, particularly the changing rates of **biological productivity** (see 152-153, 156-157).

The Earth is tilted at an angle of **23.5°** relative to its orbit around the sun (**plane of the ecliptic**). This axial tilt results in the succession of seasons as the distribution of most intense sunlight and solar insolation shifts between the Northern and Southern Hemispheres. For example, when it is summer in the Northern Hemisphere, it is winter in the Southern Hemisphere *(and vise versa).*

At the **solstice**, the sun is directly over 23.5°N (**Tropic of Cancer**, approximately June 21, Northern Hemisphere summer) or 23.5°S (**Tropic of Capricorn**, approximately December 21, Southern Hemisphere summer)

> *note:* at the time of the June 21 solstice, no sunlight strikes the Earth's surface south of 66.5° S (*Antarctic Circle*) while the region north of 66.5° N (*Arctic Circle*) experiences 24 hours of daylight. Just the reverse occurs during the December 21 solstice (i.e., total darkness north of the Arctic Circle and continuous sunlight south of the Antarctic Circle).

At the **equinox**, the sun is directly over the equator, 0° (approximately March 21 and approximately September 21)

> *note:* at the time of the equinox, the length of day and night is equal (12 hours) everywhere across the surface of the Earth.

Earth's orbit around the Sun is not a perfect circle. In detail, the orbit is slightly elliptical. Earth is closest to the Sun on ~January 3 (147×10^6 km), called **perihelion**, and is furthest on ~July 6 (152×10^6 km), called **aphelion**. The difference between perihelion and aphelion (5×10^6 km) is very small relative to the greatest diameter of Earth's orbital path (~299×10^6 km). **Eccentricity** is a term to describe the degree of deviation from a perfect circle; the greater the eccentricity, the greater the elliptical deviation from a circle. A perfect circle has an eccentricity of 0 and a flattened circle (= straight line) has an eccentricity of 1. Earth's orbit around the Sun has an eccentricity of only 0.17. You'll notice that perihelion falls during Northern Hemisphere winter, and aphelion during summer. Earth's eccentricity has a minimal effect on yearly changes in insolation across latitude and has little to do with the progression of the seasons.

Equatorial view of Earth's orbital plane

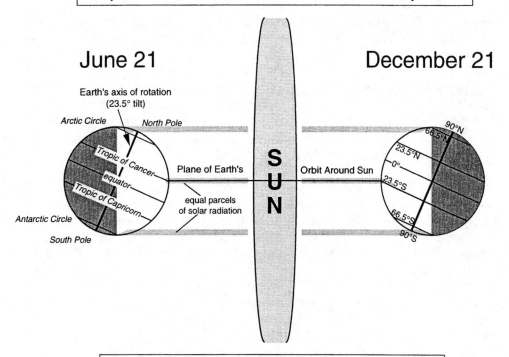

June 21

Earth's axis of rotation (23.5° tilt)

Arctic Circle — North Pole

Tropic of Cancer

equator

Tropic of Capricorn

Antarctic Circle

South Pole

Plane of Earth's

equal parcels of solar radiation

S U N

Orbit Around Sun

December 21

90°N
66.5°N
23.5°N
0°
23.5°S
66.5°S
90°S

Polar view of Earth's orbital plane

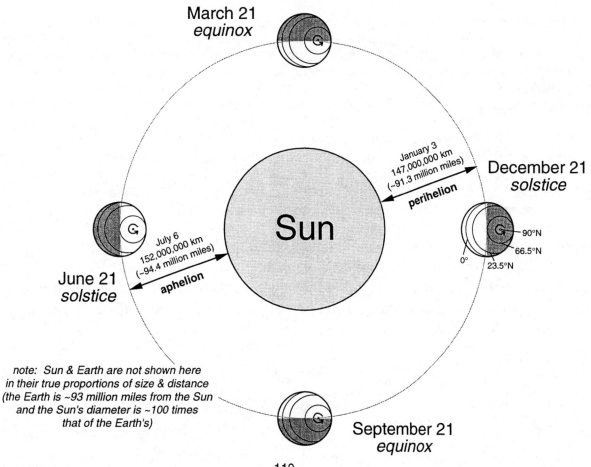

March 21
equinox

January 3
147,000,000 km
(~91.3 million miles)
perihelion

December 21
solstice

90°N
66.5°N
0°
23.5°N

Sun

July 6
152,000,000 km
(~94.4 million miles)
aphelion

June 21
solstice

note: Sun & Earth are not shown here
in their true proportions of size & distance
(the Earth is ~93 million miles from the Sun
and the Sun's diameter is ~100 times
that of the Earth's)

September 21
equinox

Seawater Density: The Role of Heat & Salt

Temperature, salinity, and pressure control the density of seawater, and the relationship among the three variables is complex. For our purposes, we will focus on the effects of temperature and salinity. The more heat liquid water absorbs, the faster the water molecules vibrate (sensible heat) and the further apart the molecules become. In this way, warm water is less dense than cold water. Therefore, warm water floats on cold water and accounts for the thermal and **density stratification** (layering) of the ocean in the tropical and temperate climate belts where there is sufficient solar radiation to warm the surface waters. In other words, warm, less dense water occurs as a surface layer over cold, dense deep waters in the low to mid-latitudes (see p. 122-123).

Substances dissolved in water also make the water more dense. Salt water is more dense than freshwater. The higher the **salinity**, the greater the density of seawater. In places where **precipitation and runoff** from the land are high, river water (freshwater) can greatly dilute coastal waters making them less salty and less dense than average seawater. An **estuary** is a place where the river meets the sea. In some situations the less dense river water may mix with the seawater, and in other places it may flow seaward above the more dense seawater (see p. 174-175). Surface water salinities are also diluted in tropical areas of the open ocean where there is abundant rainfall. In places where **evaporation** is high, such as in the subtropics, surface water salinities are typically greater than average seawater (see p. 126-127, 128-129). In the Mediterranean Sea, for example, excessive evaporation creates warm, salty waters which become dense enough to sink and flow out of the Mediterranean and into the North Atlantic where it forms an important intermediate water mass (see p. 138-139).

When water becomes more dense by cooling or by processes of the **hydrologic cycle**, such as evaporation or sea ice formation, it may sink to become a distinct intermediate or deep water mass. A **water mass** is a body of water that can be identified by its physical and chemical characteristics (temperature, salinity, density, dissolved gases, and dissolved nutrients). Intermediate and deep waters sink to their level of **neutral buoyancy** (equilibrium) below the sun-warmed surface waters (see p. 138-139).

Density is mass (quantity of matter) per unit volume. A typical unit of density is g/cm^3 (grams per cubic centimeter) or kg/m^3 (kilograms per cubic meter). Pure water (freshwater) has a density of 1.000 g/cm^3 (or 1000 kg/m^3). The density of seawater, which contains dissolved solids, varies between **1.022** and **1.028 g/cm³**. Because both salinity and temperature control density, tropical surface waters are closer to 1.022 g/cm^3 while polar surface waters and deep waters are closer to 1.028 g/cm^3. Oceanographers prefer a less cumbersome unit than g/cm^3 or kg/m^3 because of the important but subtle differences in seawater density. They use σ_t (**sigma tee**) = (density, in g/cm^3 - 1.000 g/cm^3) x 1000. Therefore, typical ocean density values range between **22.0** and **28.0** σ_t units (σ_t is based on *in situ* temperature, *in situ* salinity, and atmospheric pressure; it has not been corrected for pressure).

Density in Temperature-Salinity Space

Salinity (‰)

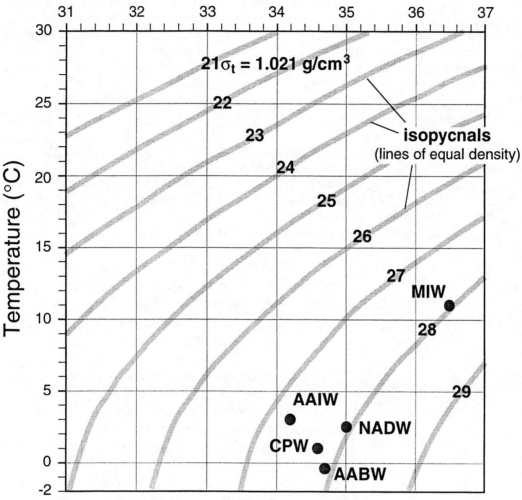

$21\sigma_t = 1.021 \text{ g/cm}^3$

isopycnals
(lines of equal density)

features of the T-S-σ_t plot
(temperature-salinity-sigma tee density)

1. the higher the σ_t value, the greater the density
2. density values increase from the upper left corner of the plot toward the lower right corner
3. the densest water masses are very cold and somewhat salty (**AABW**, **NADW**), or relatively warm and very salty (**MIW**)

Approximate Temperature and Salinity Values of Important
Deep and Intermediate Water Masses (see p.):
MIW = Mediterranean Intermediate Water (~11°C, ~36.5‰)
AAIW = Antarctic Intermediate Water (~3°C, ~34.2‰)
NADW = North Atlantic Deep Water (~2.5°C, ~35.0‰)
AABW = Antarctic Bottom Water (~-0.4°C, ~34.7‰)
CPW = Circumpolar Water (mix of NADW+AABW; ~1.0°C, 34.6‰)

Thermocline & Pycnocline: Layering of the Ocean

Solar energy heats the surface waters in the low to mid-latitudes, a result of net solar gain. This creates a warm, less dense surface layer over very cold deep waters. The **permanent thermocline** is the interval through which **temperature** decreases rapidly with increasing water depth. This interval extends from the base of the surface **mixed layer**, ~75-150 meters, to approximately 1000 m. The depth of the surface **mixed layer** is a function of mixing (homogenization) of the warmed surface waters by the day-to-day winds and storms, waves and surface currents. Winter storms tend to be bigger than summer storms; therefore the mixed layer tends to be deeper during the winter months. Summer heating causes the creation of a **seasonal thermocline** (a steeper temperature gradient than during the winter).

The entire world ocean is filled with icy cold waters below the permanent thermocline, with only small variations in temperature. A well-developed permanent thermocline exists in the low latitudes and into the mid-latitudes, although the **temperature gradient** weakens with increasing latitude. A permanent thermocline is non-existent in polar regions because surface waters are *very cold* and deep waters are *very cold*. Therefore, there is little temperature contrast/gradient between polar surface and deep waters.

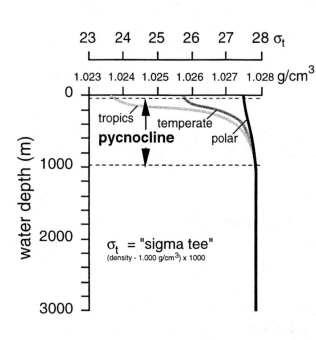

The **pycnocline** is the interval through which seawater **density** (g/cm^3) increases rapidly with increasing water depth. Oceanographers use σ_t units to describe seawater density. The pycnocline forms a **stable density barrier** between surface and deep water masses. Because density is controlled by temperature and salinity, *the permanent thermocline closely approximates the pycnocline*. Like the permanent thermocline, the top of the pycnocline is ~50-150 meters (this depth corresponds with the base of the surface mixed layer) and the base of the pycnocline is ~1000 meters in the low to mid-latitudes. Density varies little beneath the pycnocline.

Relatively limited solar heating in the high latitudes inhibits the development of a permanent thermocline and stable pycnocline. Because of the absence of stable density stratification, the high latitudes are the doorways to the deep waters of the world ocean. Deep and bottom waters, those water masses below the permanent thermocline and pycnocline, originate as polar or subpolar surface waters. During the long winter months, intense cooling and sea ice formation cause the high latitude surface waters to become denser. These dense waters sink and flow equatorward beneath the permanent thermocline and pycnocline (see p. 138-139).

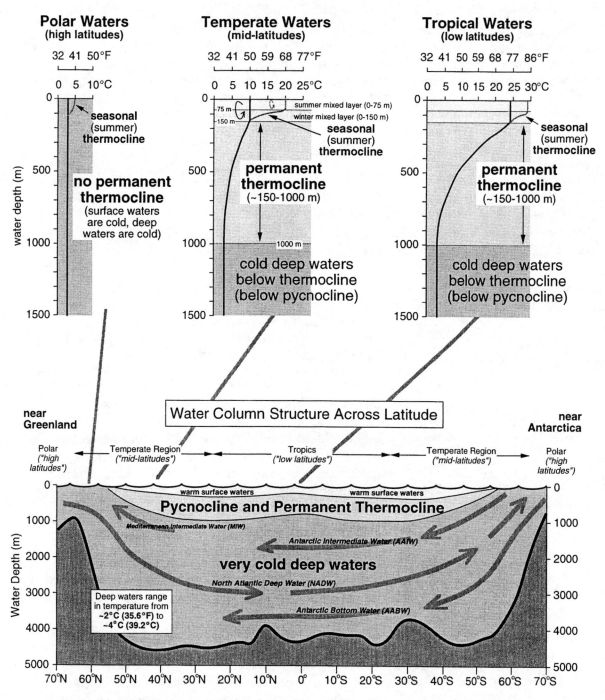

Temperature Profiles Across Latitude

Polar Waters
(high latitudes)

32 41 50°F

0 5 10°C

water depth (m)

seasonal (summer) thermocline

no permanent thermocline
(surface waters are cold, deep waters are cold)

Temperate Waters
(mid-latitudes)

32 41 50 59 68 77°F

0 5 10 15 20 25°C

-75 m
-150 m

summer mixed layer (0-75 m)
winter mixed layer (0-150 m)

seasonal (summer) thermocline

permanent thermocline
(~150-1000 m)

1000 m

cold deep waters below thermocline (below pycnocline)

Tropical Waters
(low latitudes)

32 41 50 59 68 77 86°F

0 5 10 15 20 25 30°C

seasonal (summer) thermocline

permanent thermocline
(~150-1000 m)

cold deep waters below thermocline (below pycnocline)

Water Column Structure Across Latitude

near Greenland

near Antarctica

Polar ("high latitudes") ← Temperate Region ("mid-latitudes") → ← Tropics ("low latitudes") → ← Temperate Region ("mid-latitudes") → Polar ("high latitudes")

Water Depth (m)

warm surface waters   warm surface waters

Pycnocline and Permanent Thermocline

Mediterranean Intermediate Water (MIW)

Antarctic Intermediate Water (AAIW)

very cold deep waters

North Atlantic Deep Water (NADW)

Antarctic Bottom Water (AABW)

Deep waters range in temperature from ~2°C (35.6°F) to ~4°C (39.2°C)

70°N 60°N 50°N 40°N 30°N 20°N 10°N 0° 10°S 20°S 30°S 40°S 50°S 60°S 70°S

The world ocean is **well-stratified** (layered) in the low to mid-latitudes due to solar heating of surface waters and the formation of a **permanent thermocline**. The thermocline separates warm (less dense) surface waters from icy cold (much more dense) deep waters. The **pycnocline** represents this density barrier between surface and deep waters. Insufficient solar heating in the high latitudes inhibits the formation of a permanent thermocline and pycnocline.

Earth's Rotation & the Coriolis Effect

Four aircraft take-off from 40°N, 90°W (north of St. Louis) and four aircraft take-off from 40°S, 90°W (west of Chile in the southeast Pacific), each set of aircraft with headings of due north, east, south, and west. As the aircraft fly in their straight initial bearings, the Earth rotates beneath them. In addition, each aircraft has <u>a component of velocity to the east</u> that represents the velocity of Earth's rotation at the point of take-off. This **angular velocity** at Earth's surface varies across latitude: it is faster in the lower latitudes (tropical belt), slower in the higher latitudes (temperate and polar belts):

- 1600 km/hr (994 mi/hr) at the equator (0°)
- 1400 km/hr (869 mi/hr) at 30°N&S
- 800 km/hr (497 mi/hr) at 60°N&S
- 0 km/hr at the poles (90°N&S)

If the pilots do not correct for the Earth's rotation, then the aircraft will head in a direction to the right of their original direction in the Northern Hemisphere, and to the left in the Southern Hemisphere. This <u>apparent deflection</u> is called the **Coriolis effect**. <u>The Coriolis effect influences all freely moving objects, including the motion of fluids such as air masses and water masses</u>.

To see how angular velocity, and hence Coriolis deflection, differs across latitude, compare angular velocities between latitudes:

- 0° to 30°: 1600 km/hr - 1400 km/hr = 200 km/hr difference,
- 30° to 60°: 1400 km/hr - 800 km/hr = 600 km/hr difference
- 60° to 90°: 800 km/hr - 0 km/hr = 800 km/hr difference

An air mass or water mass moving across latitude in the tropics (low latitudes) experiences less difference, and therefore less Coriolis deflection, than in the mid- to high latitudes. The Coriolis effect is zero at the equator where it changes sign from deflection to the right in the Northern Hemisphere to deflection to the left in the Southern Hemisphere (see 126-127).

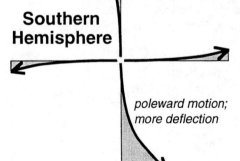

West   *East*

poleward motion; more deflection

Northern Hemisphere

equatorward motion, less deflection

Southern Hemisphere

poleward motion; more deflection

The cartoons at the left represent a schematic summary of the flight paths of the eight aircraft on the facing page. You can also think of these curved paths as the influence of the Coriolis effect on the flow of fluids on the planet, specifically the motion of air masses and water masses.

Notice two important facts: **1)** the deflection has the opposite sense on either side of the equator, and **2)** there is greater deflection towards higher latitudes.

The **Coriolis effect** is caused by the Earth's rotation; the motions of air masses and water masses (freely moving particles) are **deflected to the right in the Northern Hemisphere** and **to the left in the Southern Hemisphere.**

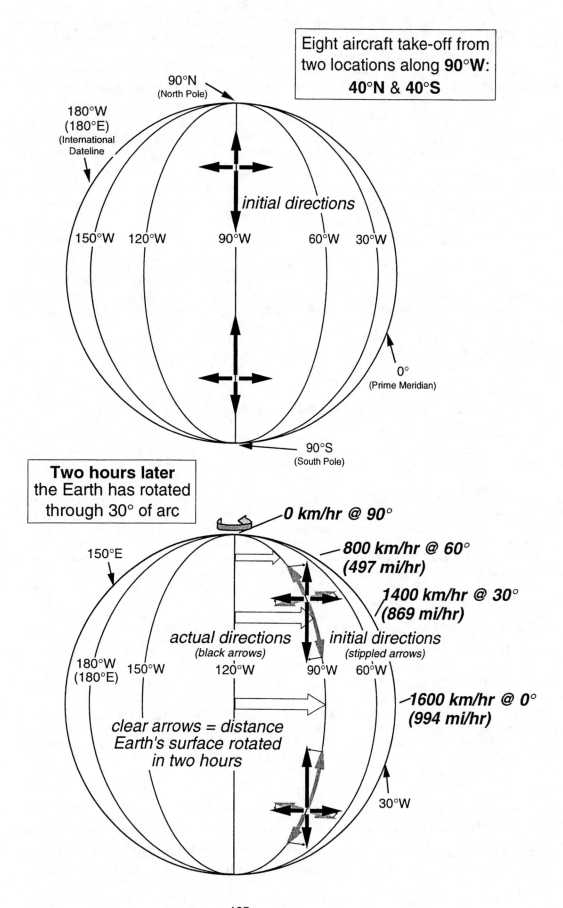

Eight aircraft take-off from
two locations along **90°W:**
40°N & 40°S

90°N
(North Pole)

180°W
(180°E)
(International
Dateline)

initial directions

150°W   120°W   90°W   60°W   30°W

0°
(Prime Meridian)

90°S
(South Pole)

Two hours later
the Earth has rotated
through 30° of arc

0 km/hr @ 90°

800 km/hr @ 60°
(497 mi/hr)

1400 km/hr @ 30°
(869 mi/hr)

150°E

actual directions
(black arrows)

initial directions
(stippled arrows)

180°W
(180°E)   150°W   120°W   90°W   60°W

1600 km/hr @ 0°
(994 mi/hr)

clear arrows = distance
Earth's surface rotated
in two hours

30°W

Prevailing Winds: Making Sense of Zonal Climate

Have you ever noticed on TV weather reports how weather patterns (zones of high and low pressure, storm tracks, and fronts) move from west to east across the United States? Have you ever noticed the general path of hurricanes in the Northern Hemisphere, first migrating west in the tropics and then gradually hooking to the right as the cyclones move away from their tropical birthplace? These phenomena are due to **zonal wind patterns** that we call the **prevailing winds**. These are not the day-to-day or hour-to-hour shifting of the local winds. Instead, these are distinct global zones of weather development and migration. Zonal refers to belts around the globe, typically between specific latitudes, like the **tropical belt**, **temperate belt**, or **polar belt**. These zonal belts are each characterized by distinct prevailing winds; storm *tracks move from east to west in the tropics and near the poles, and from west to east across the broad temperate belt.*

The atmosphere is heated most intensely by the sun in the **tropics** causing air masses to rise (= **LOW** atmospheric pressure; rising air masses exert less pressure on Earth's surface). Surface winds rush in to replace the rising air. These surface winds are called the **Northeast Trade Winds** and **Southeast Trade Winds** and they represent the prevailing winds of the tropics. The Trade Winds are among the most constant winds on the planet, except in the Pacific during an **El Nino** event (see p. 136-137). The meeting (convergence) of the NE Trades and the SE Trades near the equator is called the **Intertropical Convergence Zone (ITCZ)**. As hot, moist air masses rise in the ITCZ, they cool and water vapor condenses to produce <u>abundant precipitation</u>. Sailors sometimes refer to this region as the **"doldrums"** on account of extended periods of slack winds alternating with tropical squalls (air masses generally rise at the ITCZ, except for the downdrafts associated with thunderstorms, which provide the wind to fill the sails). The position of the ITCZ moves seasonally and is north of the equator in most regions due to the unequal distribution of land and sea between the Northern and Southern Hemisphere. The region between eastern Africa and southeast Asia experiences the greatest seasonal shifts in the position of the ITCZ. An extended period of rain associated with the seasonal movement of the ITCZ and a resultant shift in the prevailing winds is called a **monsoon** (see p. 136-137). The **tropical rain forests** are in regions of the globe traversed by the ITCZ.

Once in the upper troposphere (~16 km above Earth's surface), the air masses, now largely devoid of their moisture, diverge (split) and move poleward. As they flow north and south, they continue to cool and deflect due to the influence of the **Coriolis effect**. Between ~20° to 35°N and S latitude, the dense, dry air masses sink toward the Earth's surface (= **HIGH** atmospheric pressure). This region is at the transition between the tropical and temperate belts, and is known as the **subtropics**. These dry air masses are associated with little precipitation, and the highest rates of <u>evaporation</u> in the ocean and regions of **deserts** on land. Mariners sailing from Europe would refer to this area of the North Atlantic as the **"horse latitudes"** on account of the light winds and hot and dry conditions that resulted in longer transit times and loss of horses due to dehydration. At the surface, air masses diverge to supply the equatorward-flowing Trade Winds and the poleward-flowing **Westerlies**. The Westerlies drive the weather patterns of much of the U.S.

The **Hadley Cell** describes the conveyor-like flow of air masses in the tropics as they rise at the ITCZ, diverge and move poleward, sink in the subtropics, and are drawn back toward the ITCZ as the surface winds we call the NE and SE Trade Winds.

In the mid-latitudes, between ~35° to 60°N and S latitude, the Westerlies converge with the **Polar Easterlies**. This is where tropical air clashes with polar air along the **Polar Front**. Rising air masses create areas of low atmospheric pressure and year-round precipitation. Sailors sometimes refer to this region of the ocean as the **"roaring forties"** because of the strong winter storms that whip-up large, irregular waves. This region is characterized by highly seasonal and variable weather patterns as the Polar Front meanders north and south as a planetary-scale wave. Think of this meandering front as an unattended fire hose that is thrashing about uncontrollably (but in super slow motion!) due to the great water pressure.

Near the poles, cold dry air descends once again giving birth to the Polar Easterlies at the surface. The polar regions are characterized by little precipitation.

Notice that the Trade Winds, the Westerlies, and the Polar Easterlies are all deflected from direct north to south, or south to north pathways due to the influence of the **Coriolis Effect**.

Idealized Global Atmospheric Circulation

areas where air masses rise (**LOW** atmospheric pressure)
are characterized by **precipitation>evaporation**
areas where air masses sink (**HIGH** atmospheric pressure)
are characterized by **evaporation>precipitation**

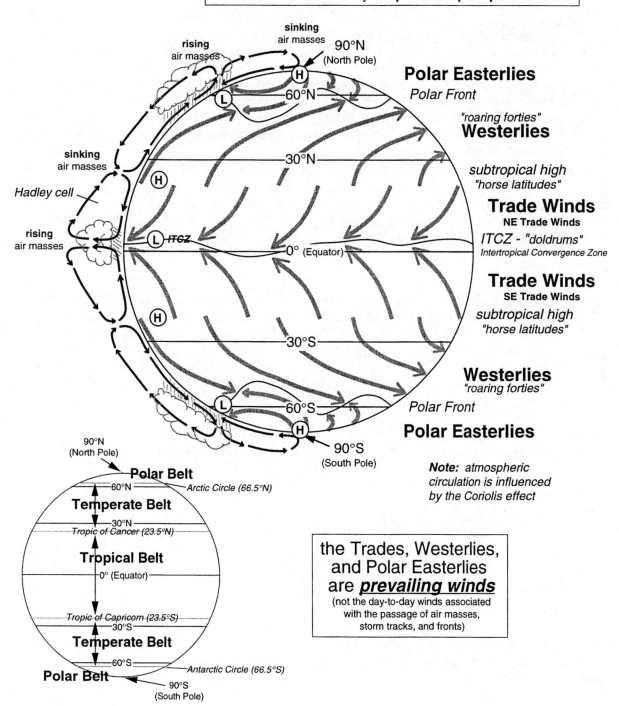

sinking air masses

rising air masses

90°N (North Pole)

60°N

Polar Easterlies
Polar Front

"roaring forties"
Westerlies

30°N

subtropical high
"horse latitudes"

Trade Winds
NE Trade Winds

ITCZ - "doldrums"
Intertropical Convergence Zone

sinking air masses

Hadley cell

rising air masses

0° (Equator)

Trade Winds
SE Trade Winds

subtropical high
"horse latitudes"

30°S

Westerlies
"roaring forties"

Polar Front

Polar Easterlies

60°S

90°S (South Pole)

Note: *atmospheric circulation is influenced by the Coriolis effect*

90°N (North Pole)

Polar Belt

60°N — *Arctic Circle (66.5°N)*

Temperate Belt

30°N — *Tropic of Cancer (23.5°N)*

Tropical Belt

0° (Equator)

Tropic of Capricorn (23.5°S)
30°S

Temperate Belt

60°S — *Antarctic Circle (66.5°S)*

Polar Belt

90°S (South Pole)

the Trades, Westerlies,
and Polar Easterlies
are ***prevailing winds***
(not the day-to-day winds associated
with the passage of air masses,
storm tracks, and fronts)

Climate & the Ocean

Equatorial Low & the Intertropical Convergence Zone (ITCZ): ~0°
- dominated by **low atmospheric pressure** (rising warm, moist air masses)
- sometimes referred to as the *"doldrums"* (hot, with light variable winds alternating with squalls)
- the position of <u>the ITCZ is north of the equator in most regions</u> and it shifts seasonally, especially in the Indian Ocean and over SE Asia resulting in monsoon rains in the summer
- <u>abundant precipitation</u> (therefore **very warm, but relatively low salinity, surface waters**)
- breeding grounds for <u>tropical cyclones</u> (hurricanes, typhoons)

> ### *Tropical Belt: 0° to ~23.5° N & S ("low latitudes")*
> - *prevailing winds = <u>Trade Winds</u> (Hadley Cell)*
> - *the most persistent winds on the planet, except during **El Nino events** when Trade Wind strength is greatly diminished in the Pacific*
> - *modest seasonality in some regions, large excursions of ITCZ in other regions produce strong seasonal changes in precipitation related to <u>monsoonal circulation</u> (dry season/rainy season)*
> - *precipitation increases towards the ITCZ (**warm to very warm surface waters**)*
> - *also an area for <u>tropical cyclone</u> development (hurricanes, typhoons)*

Subtropical High: ~20-35° N & S
- dominance by **high atmospheric pressure** (sinking dense, dry air masses); e.g., the Bermuda High
- sometimes referred to as the *"horse latitudes"* (light, sporadic winds; horses lost to dehydration)
- dry in all seasons, <u>high evaporation</u> (therefore **warm waters with the highest surface salinities** of the open ocean)

> ### *Temperate Belt: ~23.5° to ~66.5° N & S ("mid-latitudes")*
> - *prevailing winds = <u>Westerlies</u>*
> - *<u>strong seasonality</u> (related to movement of the Polar Front)*
> - *a fast moving current of air called a Jet Stream races along the steep temperature and pressure gradients where polar and tropical air masses meet*
> - *ample precipitation in all seasons (**warm to cool surface waters** with **relatively low surface salinities away from the subtropics**)*

Subpolar Low and the <u>Polar Front</u>: ~35-60° N & S
- dominated by **low atmospheric pressure** (meeting of tropical and polar air masses); e.g., the Aleutian Low and the Icelandic Low
- the Polar Front often meanders from day to day or week to week and its position shifts north-south with the seasons
- abundant precipitation associated with the front
- source of <u>extratropical cyclones</u> (especially in winter; e.g., **"Nor'easters"**)

> ### *Polar Belt: ~66.5° to 90° N & S ('high latitudes")*
> - *prevailing winds = <u>Polar Easterlies</u>*
> - *precipitation decreases towards the poles*
> - *<u>sea-ice formation</u> during winter (**cold seas**)*

Polar High: ~90° N & S ("poles")
- dominated by **high atmospheric pressure** (sinking of cold, dry air masses)
- sparse precipitation

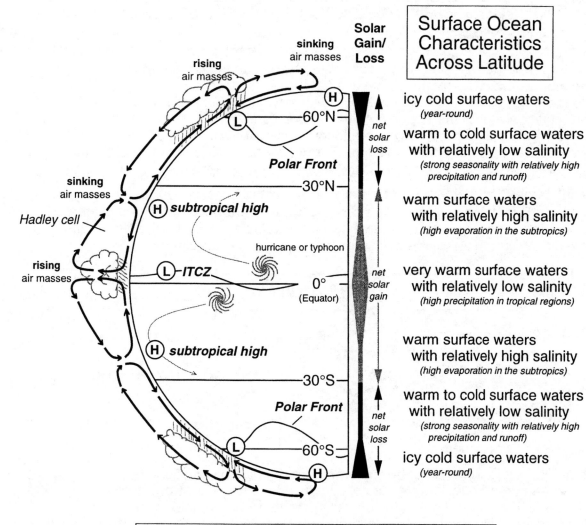

Surface Ocean Characteristics Across Latitude

Solar Gain/Loss

icy cold surface waters
(year-round)

warm to cold surface waters with relatively low salinity
(strong seasonality with relatively high precipitation and runoff)

warm surface waters with relatively high salinity
(high evaporation in the subtropics)

very warm surface waters with relatively low salinity
(high precipitation in tropical regions)

warm surface waters with relatively high salinity
(high evaporation in the subtropics)

warm to cold surface waters with relatively low salinity
(strong seasonality with relatively high precipitation and runoff)

icy cold surface waters
(year-round)

sinking air masses
rising air masses
sinking air masses
Hadley cell
rising air masses

H — 60°N
L
Polar Front
30°N
H *subtropical high*
hurricane or typhoon
L-ITCZ
0° (Equator)
H *subtropical high*
30°S
Polar Front
L — 60°S
H

net solar loss
net solar gain
net solar loss

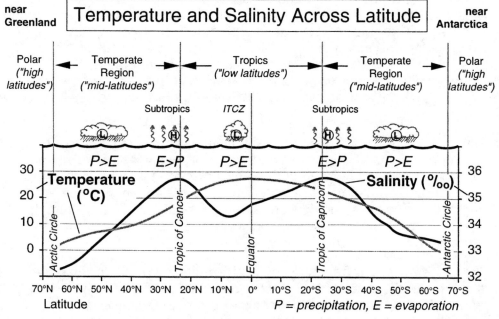

near Greenland

Temperature and Salinity Across Latitude

near Antarctica

Polar ("high latitudes") — Temperate Region ("mid-latitudes") — Tropics ("low latitudes") — Temperate Region ("mid-latitudes") — Polar ("high latitudes")

Subtropics — ITCZ — Subtropics

L — H — L — H — L

P>E — E>P — P>E — E>P — P>E

Temperature (°C)   **Salinity (‰)**

Arctic Circle — Tropic of Cancer — Equator — Tropic of Capricorn — Antarctic Circle

Latitude: 70°N 60°N 50°N 40°N 30°N 20°N 10°N 0° 10°S 20°S 30°S 40°S 50°S 60°S 70°S

Latitude

P = precipitation, E = evaporation

129

Ekman Transport & Wind-Driven Circulation: Motion of the Surface Ocean

The **prevailing winds** create a drag (wind stress) on the ocean surface, thus transferring momentum. The momentum gained at the surface is further transferred deeper into the water column, but energy is lost with increasing depth. The Coriolis effect deflects the moving water of the upper 75-150 m of the water column (see p. 124-125). A decrease in current speed, coupled with continuous deflection with increasing depth creates a theoretical spiral of water called the **Ekman spiral**. Due to the Coriolis effect, the surface current moves ~45° to the prevailing wind. Adding all the vectors (magnitude and direction) of the Ekman spiral yields a net current direction that is ~90° to the prevailing wind. This composite current is called **Ekman transport** and it controls the motion of the surface ocean. The diagram below depicts the situation for the Northern Hemisphere (the influence of the Coriolis effect is just the opposite in the Southern Hemisphere).

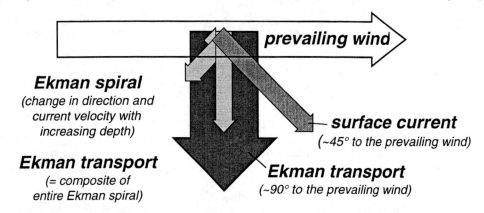

Ekman spiral
(change in direction and current velocity with increasing depth)

Ekman transport
(= composite of entire Ekman spiral)

prevailing wind

surface current
(~45° to the prevailing wind)

Ekman transport
(~90° to the prevailing wind)

Energy of the prevailing winds sets the uppermost water column in motion. This movement of the upper water masses is called **wind-driven circulation**, and the motion is in a direction to the right of the prevailing winds in the Northern Hemisphere and to the left of the prevailing winds in the Southern Hemisphere. Ekman transport causes near-surface waters to **converge** (pile-up) in subtropical regions thereby creating subtle "hills", "domes", or "ridges" on the ocean surface, and it causes waters to **diverge** (move apart) in subpolar regions and along the equator, creating "depressions" or "valleys". Divergence of water masses causes **upwelling** of deeper waters (see p. 134-135). These subtle highs and lows on the ocean surface are not visible because the relief is less than 2 meters (<6.6 ft.) higher or lower than the average level of the sea over broad areas of the ocean.

Gyres are the large horizontal wind-driven current systems that circulate around the subtle domes and depressions on the ocean surface. For example, the **subtropical gyres** represent large circulation cells around the hills created by convergence in the subtropics. These currents transport warm waters poleward along the western sides of the ocean basins and cool waters equatorward along the eastern sides. Subtropical gyres are prominent features of circulation in the North and South Atlantic, North and South Pacific, and the Indian Ocean.

Circulation of the **upper water masses** (= wind-driven circulation) is *independent* of the circulation of the **intermediate and deep water masses** (= thermohaline circulation) because the **pycnocline** (or permanent thermocline) provides a *stable density barrier* between the less dense (warmer) near-surface waters and the more dense (very cold) deep waters (see p. 122-123).

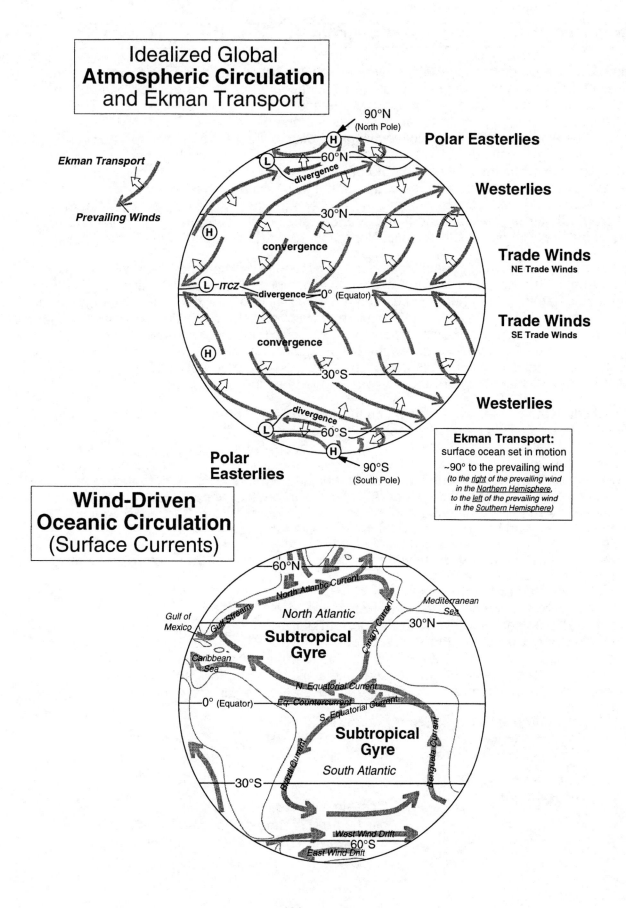

Idealized Global
Atmospheric Circulation
and Ekman Transport

Ekman Transport

Prevailing Winds

90°N
(North Pole)

Polar Easterlies

60°N

divergence

Westerlies

30°N

convergence

Trade Winds
NE Trade Winds

ITCZ

divergence — 0° (Equator)

Trade Winds
SE Trade Winds

convergence

30°S

Westerlies

divergence
60°S

90°S
(South Pole)

**Polar
Easterlies**

Ekman Transport:
surface ocean set in motion
~90° to the prevailing wind
*(to the right of the prevailing wind
in the Northern Hemisphere,
to the left of the prevailing wind
in the Southern Hemisphere)*

**Wind-Driven
Oceanic Circulation**
(Surface Currents)

60°N

North Atlantic Current

Gulf of
Mexico

Gulf Stream

Mediterranean
Sea

North Atlantic

Canary Current

30°N

**Subtropical
Gyre**

Caribbean
Sea

N. Equatorial Current

Eq. Countercurrent

S. Equatorial Current

0° (Equator)

Brazil Current

**Subtropical
Gyre**

Benguela Current

South Atlantic

30°S

West Wind Drift

60°S

East Wind Drift

Geostrophic Currents: Coriolis & Gravity in Balance

The prevailing winds provide the energy to drive the surface currents of the world ocean. Ekman transport and the Coriolis effect cause surface waters to **converge** ("pile-up") in the subtropics and **diverge** (move apart) at the equator and in subpolar waters. This creates subtle "hills" and "valleys" on the ocean surface of < 2 meters (<6.6 ft.). Gravity acts on the water to pull it downslope while the Coriolis effect (Ekman transport) is working in the opposite direction (to the right in the Northern Hemisphere)(see p. 130-131). **Geostrophic flow** represents the partial balance between gravity and Coriolis. **Geostrophic currents** are surface currents that flow around these subtle domes and depressions, and most surface currents in the ocean are geostrophic. For example, the **subtropical gyres** represent the flow of upper water masses around slightly elevated domes in the North and South Atlantic, North and South Pacific, and in the Indian Ocean.

Earth's rotation from west to east, compounded by the typically strong Trade Winds in the tropics, cause the waters to "pile-up" on the western sides of the ocean basins. Notice, for example, what happens when you push on a pan or pail of water; the water rises on the side you pushed from. As the Earth rotates from west to east, the continents "push" on the western sides of the ocean basins ("pans") and therefore, the ocean surface stands slightly higher on the western sides of the ocean basins. In this way, the dome of water created by convergence in the subtropics is not located in the center of the ocean basin, but is instead displaced toward the western side (see facing page).

This offset dome causes water flowing on the western side of the dome to flow faster than the eastern side. This is called **"western intensification"**. It's like taking a garden hose and partially constricting the flow; the water shoots out farther and faster. The same is true for the upper water masses; the surface currents are forced through a narrower passage between the continents and the crest of the dome causing them to flow faster. For example, in the profile of the North Atlantic shown on the facing page, notice that the distance between North America and the crest of the dome in the **Sargasso Sea** is much narrower than the distance between the crest of the dome and Africa.

Strong **western boundary currents** mark the western sides of the subtropical gyres:
- **Gulf Stream** in the North Atlantic
- **Brazil Current** in the South Atlantic
- **Kuroshio Current** in the North Pacific
- **East Australian Current** in the South Pacific
- **Agulhas Current** in the Indian Ocean

These waters play an important role in transporting tropical heat to the cooler high latitudes. For example, the **Gulf Stream** ultimately becomes the **North Atlantic Current** (also called North Atlantic Drift) and moderates the climate of northern Europe.

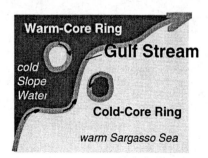

The flow of the **Gulf Stream** is like a wide meandering river, although its course is not nearly as rigidly constrained as the banks of floodplain sediments that border a river. Occasionally, the meander loops become pinched-off creating eddies of rotating water called **warm-core rings** and **cold-core rings.**

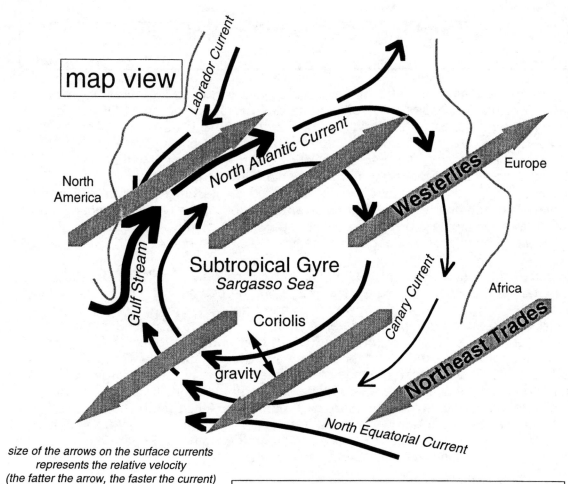

map view

Labrador Current

North Atlantic Current

North America

Europe

Westerlies

Gulf Stream

Subtropical Gyre
Sargasso Sea

Coriolis

Africa

Canary Current

gravity

Northeast Trades

North Equatorial Current

size of the arrows on the surface currents represents the relative velocity (the fatter the arrow, the faster the current)

convergence of surface waters creates a dome of water in the subtropics;
rotation of the Earth causes the dome to be displaced to the west;
the **subtropical gyre** is the motion of upper waters around the dome;
the subtropical gyre isolates a portion of the North Atlantic called the **Sargasso Sea**

profile

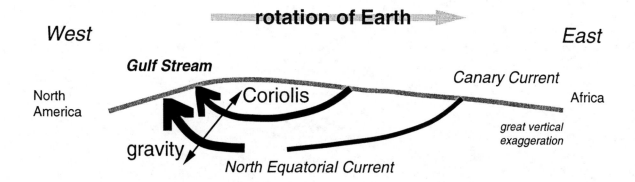

rotation of Earth

West

East

Gulf Stream

North America

Coriolis

Canary Current

Africa

gravity

great vertical exaggeration

North Equatorial Current

Wind-Driven Upwelling: Gateway to Biological Productivity

The **wind-driven surface currents** of the world ocean, including the gyres, represent *horizontal flow of water masses*. **Upwelling** and **downwelling** represent the *vertical movement of water masses* in the world ocean. **Downwelling** is associated with the sinking of cold, dense water in the high latitudes of the Atlantic to form North Atlantic Deep Water and Antarctic Bottom Water, and the sinking of warm, salty water in the Mediterranean Sea to form Mediterranean Intermediate Water (see p. 138-139). **Upwelling** is the upward movement of water due to the displacement of surface waters by the prevailing winds and resultant **Ekman transport**, or by the Coriolis effect acting on surface currents to create a divergence of surface waters. There are two principal types of upwelling: coastal and oceanic divergence

Coastal upwelling occurs when prevailing winds blow roughly parallel to the shore and Ekman transport pushes surface waters away from the coast (facing page shows an example from the Northern Hemisphere where Ekman transport is to the right of the prevailing wind). Deeper, nutrient-rich waters come up to replace displaced surface waters. This results in **high biological productivity** by primary producers (plankton) which in turn supports a greater abundance of fish and other marine life. Coastal upwelling is prevalent on the eastern sides of the ocean basins (western margins of the continents). In many areas, the strength of upwelling varies seasonally. Examples of seasonal coastal upwelling include the margin off northwest Africa (Morocco, Spanish Sahara, Mauritania, Senegal) and west Africa (Namibia, Angola) in the eastern Atlantic, off California and Peru in the eastern Pacific, and along the Somalia, South Yemen, and Oman margins (summer) in the northwest Indian Ocean.

Oceanic divergence also results in the upwelling of deeper, nutrient-rich waters. The resulting elevated rates of primary productivity support larger communities of animals. Divergence is created where *prevailing winds converge* but the resulting Ekman transport causes the *near-surface waters to diverge*. This occurs in subpolar waters of the North Atlantic and North Pacific, and around the continent of Antarctica (see p. 130-131). These are rich feeding grounds for a number of whale species. Upwelling also occurs along the equator where the sign of the Coriolis changes direction from Ekman transport to the right of the prevailing wind (Northern Hemisphere) to left of the prevailing wind (Southern Hemisphere). This latter example is shown on the facing page and is referred to as **equatorial upwelling**. The Galapagos Islands, famous for the research of **Charles Darwin** and his evolutionary theory of natural selection, are located on the equator in the eastern Pacific and are characterized by the diverse and abundant marine life that is supported by equatorial upwelling.

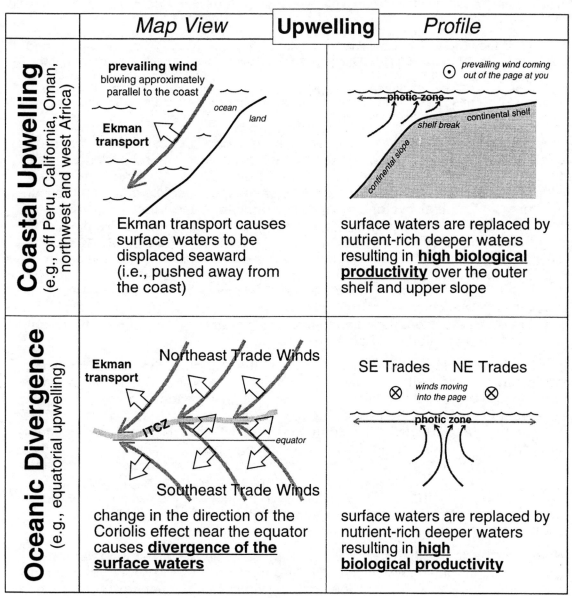

	Map View	**Upwelling**	*Profile*

Coastal Upwelling (e.g., off Peru, California, Oman, northwest and west Africa)

prevailing wind blowing approximately parallel to the coast

ocean

land

Ekman transport

Ekman transport causes surface waters to be displaced seaward (i.e., pushed away from the coast)

⊙ prevailing wind coming out of the page at you

photic zone

continental shelf

shelf break

continental slope

surface waters are replaced by nutrient-rich deeper waters resulting in **high biological productivity** over the outer shelf and upper slope

Oceanic Divergence (e.g., equatorial upwelling)

Ekman transport

Northeast Trade Winds

ITCZ

equator

Southeast Trade Winds

change in the direction of the Coriolis effect near the equator causes **divergence of the surface waters**

SE Trades   NE Trades

⊗ winds moving into the page ⊗

photic zone

surface waters are replaced by nutrient-rich deeper waters resulting in **high biological productivity**

convergence of the Trade Winds near the equator defines the Intertropical Convergence Zone (ITCZ)

Ekman Transport:
surface ocean set in motion
~90° to the prevailing wind
(to the *right* in the Northern Hemisphere,
to the *left* in the Southern Hemisphere)

Climate Variability: Cyclones, Monsoons & El Nino

Have you ever noticed how some summers seem hotter and drier than others, or that some winters produce much more precipitation than "average"? **Weather** refers to the day-to-day changes in our atmosphere, while **climate** describes the long-term patterns of weather for each region of the Earth. Different climate zones (e.g., tropical, subtropical, arid, semi-arid, mediterranean, temperate, northern temperate, mountain, coastal, polar) experience a characteristic progression of weather patterns and seasonal extremes of temperature and precipitation as the amount of solar radiation increases and decreases during the course of a year. But no two years are exactly alike. **Interannual variability** relates to perturbations of the "average" or expected weather for a given season of the year. If we are to fully understand our ocean-climate system, we must understand the natural extremes of climate from one region to another and from one year to the next.

Storms are an important part of seasonal weather. A **cyclone** is a major storm that grows around a **low pressure** area. The winds blow counterclockwise around cyclones in the Northern Hemisphere and clockwise in the Southern Hemisphere. Cyclones are driven by the prevailing winds and steered by the Coriolis effect and other low and high pressure cells in their paths. **Tropical cyclones** (called **hurricanes** in the Western Hemisphere and **typhoons** in the Eastern Hemisphere) represent safety valves for the release of excess heat that builds up every year in the tropics and subtropics (see p. 128-129). These powerful seasonal storms transport much of this excess heat towards the cooler high latitudes. **Extratropical cyclones** are storms that are born outside of the tropics. During the winter months, the position of the **Polar Front** moves into the temperate belt. Powerful winter storms such as the **"Nor'easters"** that batter the eastern seaboard of the U.S. are generated by the intensified temperature gradient across the Polar Front (clash of relatively warm air and bitterly cold arctic air).

Monsoon circulation refers to seasonal changes in the pattern of the prevailing winds. In a simplistic way, it is similar to the change from an onshore breeze during the daylight hours to an offshore breeze at night along the coast (see facing page). In tropical regions, this phenomenon is related to the yearly migration of the **Intertropical Convergence Zone** (ITCZ) as the "heat equator" shifts north during the Northern Hemisphere summer and south during the Northern Hemisphere winter. Mountainous regions at the edge of the tropics, such as the southwest U.S., also experience a "summer monsoon". The term "monsoon" is commonly applied to the season that experiences increased precipitation. The seasonal movement of the ITCZ is particularly pronounced in the region of the Indian Ocean: from east Africa to southeast Asia. Places like India and the Philippines experience months of torrential rains followed by months with little rain.

The **El Nino Southern Oscillation** (ENSO) is a prime example of natural interannual variability in the ocean-climate system. An **El Nino event** occurs every 3-7 years and it can greatly perturb the "expected" seasonal progression in weather patterns across the globe. During a "normal" year, there is a strong atmospheric pressure gradient between the Southeast Pacific HIGH and the Indo-Australian LOW. This results in strong Trade Winds which cause warm water to pile-up (thicken) in the western Pacific due to the relatively constricted flow through the passages between Australia and Indonesia. The resultant thick layer of tropical water is called the **Western Pacific Warm Pool**. However, during an El Nino event, the atmospheric pressure gradient across the tropical Pacific weakens, Trade Wind strength slackens, and warm water caps the normally cool, nutrient-rich waters that upwell along the equatorial divergence and continental margin of Peru and Ecuador. The warm current shows up around Christmas time and results in fish kills, flooding rains, and economic disaster for the local people. The effects of a severe El Nino event are also felt beyond the eastern tropical Pacific, including widespread droughts in some regions, floods in others, severe winter storms along the California coast, and milder winters in the northeast U.S.

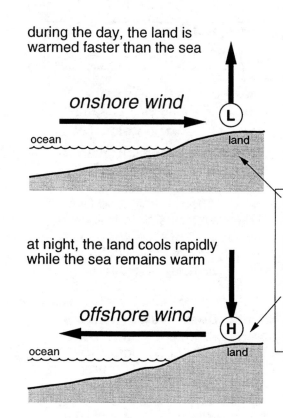

during the day, the land is warmed faster than the sea

onshore wind

ocean   land

at night, the land cools rapidly while the sea remains warm

offshore wind

ocean   land

Monsoon Circulation

Temperature contrasts between the land and the ocean produce small-scale <u>daily shifts in the winds</u> (***coastal sea breeze***), as well as large-scale <u>seasonal shifts in regional weather patterns</u> (***monsoons***).

Summer heating causes air masses to rise over east Africa, southern and southeastern Asia as the **ITCZ shifts north**. This draws in moist air from the Indian Ocean which results in **very high precipitation** over land.

Winter cooling over the Himalaya and Tibetan Plateau causes air masses to sink. The **ITCZ shifts south**, the winds reverse, and the rains end.

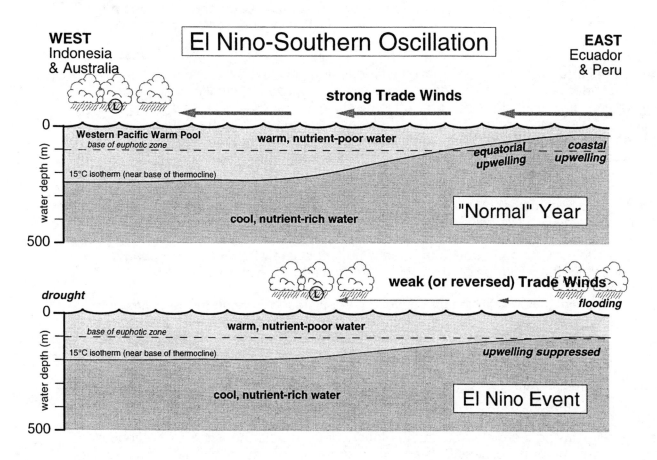

El Nino-Southern Oscillation

WEST
Indonesia & Australia

EAST
Ecuador & Peru

strong Trade Winds

Western Pacific Warm Pool
base of euphotic zone

warm, nutrient-poor water

equatorial upwelling   *coastal upwelling*

15°C isotherm (near base of thermocline)

cool, nutrient-rich water

"Normal" Year

water depth (m)   0   500

weak (or reversed) Trade Winds

drought

flooding

base of euphotic zone

warm, nutrient-poor water

15°C isotherm (near base of thermocline)

upwelling suppressed

cool, nutrient-rich water

El Nino Event

water depth (m)   0   500

Density-Driven Downwelling & Thermohaline Circulation: Motion of the Deep Ocean

Intense **winter cooling** and **sea-ice formation** in polar seas during the several months of darkness results in the formation of cold, dense waters. These sink and move equatorward beneath the pycnocline as "deep" or "bottom" waters. Icy cold waters (<4°C, or <40°F) fill the entire world ocean below the pycnocline. **Thermohaline circulation** is the **density-driven circulation** of the deep ocean. Thermohaline circulation also includes the production and circulation of "intermediate" water masses. The term "thermohaline" implies that both temperature and salinity are important in the production of deep and intermediate water masses. Intermediate and deep waters sink to their level of **neutral buoyancy** (density equilibrium) below the solar-warmed surface waters. There is a vertical component of thermohaline circulation related to the **downwelling** of dense waters, but much of the motion is related to horizontal **advection** (flow) through the ocean basins of the world.

Deep waters form only in the Atlantic today. **North Atlantic Deep Water** (**NADW**) forms in the northern North Atlantic (Greenland-Norwegian Sea) and **Antarctic Bottom Water** (**AABW**) forms in the southern South Atlantic (Weddell Sea of Antarctica). The deep waters of the Indian and Pacific are a mix of NADW and AABW called **Circumpolar Water** (sometimes called Common Water).

Deep water masses, like surface currents, are influenced by the **Coriolis effect** resulting in strong **western boundary currents**. As deep waters flow equatorward beneath the pycnocline they hug the lower continental slopes and rises on the western sides of the ocean basins.

Other waters also form in the mid- to high latitudes and sink to intermediate depths. These **intermediate waters** are less dense, or are produced in smaller volume than deep waters. For example, **Antarctic Intermediate Water** (**AAIW**) forms at the Antarctic Convergence (Polar Front), then sinks and flows north into the South Atlantic, Caribbean Sea, Gulf of Mexico and North Atlantic. Similar intermediate waters also form in the North and South Pacific. **Mediterranean Intermediate Water** (**MIW**) is a warm, salty water mass produced in the Mediterranean Sea due to high evaporation rates in the subtropics. Although MIW is among the most dense waters formed in the ocean today it doesn't sink to become deep or bottom water because the relatively small volume of water that spills over the Gibraltar sill into the North Atlantic mixes with other water masses to dilute its originally dense character (see p. 120-121). Therefore, MIW sinks to about 1000 m and turns toward the north due to the Coriolis effect where it provides an important source of salt for the winter production of North Atlantic Deep Water.

Thermohaline circulation is important because it ventilates (oxygenates) the entire water column of the world ocean below the pycnocline. The motion of intermediate, deep, and bottom waters redistributes heat, salt, dissolved gases, and nutrients. Thermohaline circulation is intimately linked with wind-driven circulation above the pycnocline to create a **"global conveyor"** of world ocean circulation (see p. 140-141). The seasonal formation of sea-ice releases latent heat to the atmosphere in the high latitudes. Salt supplied by intermediate waters like MIW, or by western boundary currents such as the Gulf Stream help the surface waters of the northern North Atlantic sink to become NADW. In addition to playing the major role in the global conveyor, thermohaline circulation also carries away some of the excess CO_2 and other greenhouse gases that have been rapidly building-up in our atmosphere due to the burning of fossil fuels.

Thermohaline Circulation

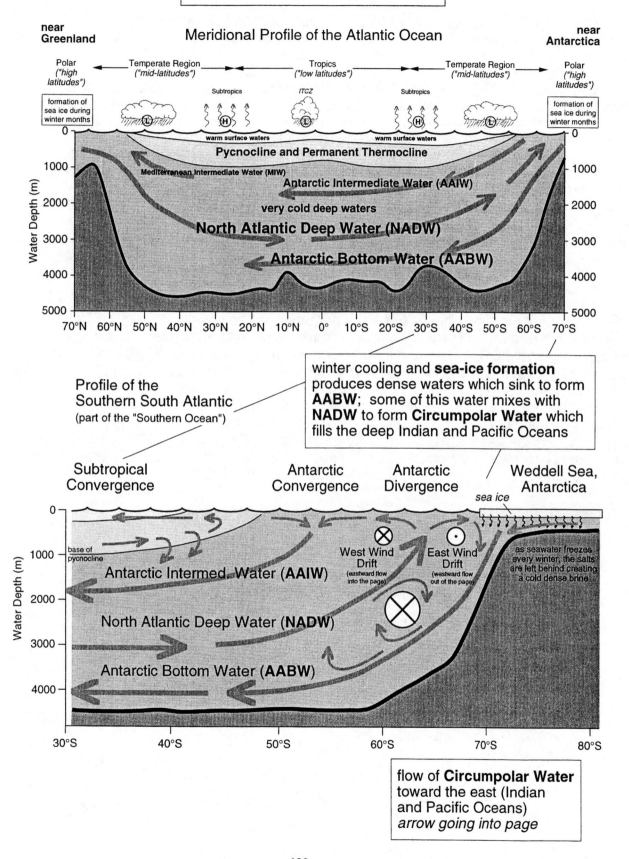

Meridional Profile of the Atlantic Ocean

near Greenland

near Antarctica

Polar ("high latitudes")

Temperate Region ("mid-latitudes")

Tropics ("low latitudes")

Temperate Region ("mid-latitudes")

Polar ("high latitudes")

Subtropics   ITCZ   Subtropics

formation of sea ice during winter months

formation of sea ice during winter months

warm surface waters   warm surface waters

Pycnocline and Permanent Thermocline

Mediterranean Intermediate Water (MIW)

Antarctic Intermediate Water (AAIW)

very cold deep waters

North Atlantic Deep Water (NADW)

Antarctic Bottom Water (AABW)

Water Depth (m): 0, 1000, 2000, 3000, 4000, 5000

70°N 60°N 50°N 40°N 30°N 20°N 10°N 0° 10°S 20°S 30°S 40°S 50°S 60°S 70°S

winter cooling and **sea-ice formation** produces dense waters which sink to form **AABW**; some of this water mixes with **NADW** to form **Circumpolar Water** which fills the deep Indian and Pacific Oceans

Profile of the Southern South Atlantic (part of the "Southern Ocean")

Subtropical Convergence

Antarctic Convergence

Antarctic Divergence

Weddell Sea, Antarctica

sea ice

base of pycnocline

Antarctic Intermed. Water (AAIW)

West Wind Drift (eastward flow into the page)

East Wind Drift (westward flow out of the page)

as seawater freezes every winter, the salts are left behind creating a cold dense brine

North Atlantic Deep Water (NADW)

Antarctic Bottom Water (AABW)

Water Depth (m): 0, 1000, 2000, 3000, 4000

30°S 40°S 50°S 60°S 70°S 80°S

flow of **Circumpolar Water** toward the east (Indian and Pacific Oceans) *arrow going into page*

The Global Conveyor: World Ocean Circulation

The **global conveyor** describes the complete circuit of global ocean circulation involving **horizontal flow** of surface and deep waters, and the **vertical flow** of downwelling and upwelling. Waters of the thermocline and surface mixed layer move under the influence of wind-driven circulation; waters below the pycnocline (permanent thermocline) move by density-driven thermohaline circulation. Surface waters move independently of deep and intermediate waters because of the pycnocline that separates them. *Downwelling and upwelling provide the processes, which link the surface and deep ocean.*

Surface waters cool and sink in the Atlantic and spread to fill the world ocean with cold deep waters. Upwelling at the margins of the ocean, around the continent of Antarctica (Antarctic Divergence), and in the equatorial Pacific return waters to the surface. Surface currents then return these waters back to the Atlantic. There is a **net export of deep water out of the Atlantic** and **net import of surface water into the Atlantic**.

Deep and intermediate water masses acquire their physical and chemical characteristics at the surface where they formed (see p. 138-139). These properties include **temperature**, **salinity**, and **density**, which are considered **conservative properties** because they are modified only by **diffusion** or **mixing** with other water masses. Other properties acquired at the surface include **nutrients** and **dissolved gases** (oxygen, O_2, and carbon dioxide, CO_2). These are **non-conservative properties** because they are modified by **biochemical cycles** (see p. 154-155). For example, because cold water holds more dissolved gasses than warm water, deep waters are initially oxygen-rich. However, these waters lose O_2 over time due to animal respiration and the decomposition of organic matter (both of which yield CO_2). The older the water mass, the longer it has been away from the surface, the lower its O_2 content and the higher its CO_2 content. The oldest deep waters are in the North Pacific. These waters have been away from the surface for 500-1000 years and they are rich in CO_2 and nutrients.

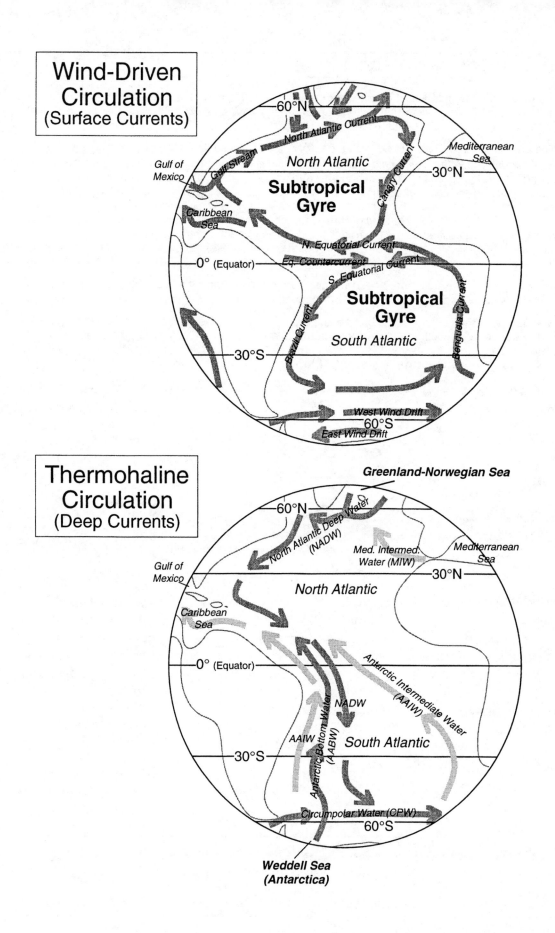

Waves: Energy on the Move

Waves represent the transmission of **energy,** *not* **mass** along the interface between fluids of differing density; for example, between the surface of the ocean and the atmosphere. Because the waves do not involve mass transfer, they are not influenced by the Coriolis effect.

Wind waves are produced by the day to day changes in weather, particularly **storms** *(not the prevailing winds, which impart momentum to the upper water masses and drive fluid motion).* Storm waves move across the surface ocean as **swell.** Because long wavelength waves travel faster than short wavelength waves, the trains of waves become sorted by wavelength with the longer waves racing away from the storm that generated them faster than the shorter waves. The sorting of waves is called **dispersion.** Wave trains originating from different directions can pass through each other creating **interference patterns** in the swell. Waves become more irregular as the troughs and crests of each contributing wave train add to or subtract from each others waveform (constructive or destructive interference). Occasionally a very tall wave is produced by the additive effect of wave crests from multiple wave trains creating a short-lived **rogue wave.**

Waves also occur below the ocean surface. **Internal waves** move along the interface of subsurface waters of differing density such as along the pycnocline (thermocline).

Waves eventually lose their energy in the surf zone. Kinetic energy of the waves is transferred to the ocean floor when the waves "touch bottom" in intermediate (<L/2) and shallow water (<L/20): the *seafloor impedes the wave motion and causes the waves to slow down and sediment is put into motion.* Wave energy in coastal waters sets up **longshore currents** and resultant **longshore drift** of sediment (see p. 170-171)

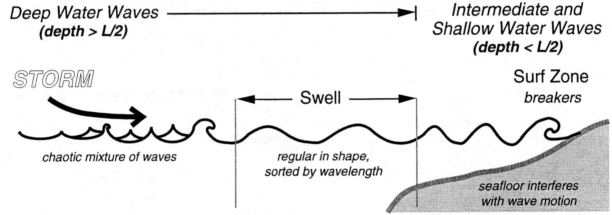

Deep Water Waves
(depth > L/2)

Intermediate and Shallow Water Waves
(depth < L/2)

STORM

Surf Zone
breakers

← Swell →

chaotic mixture of waves

regular in shape, sorted by wavelength

seafloor interferes with wave motion

Storm Waves ("Sea")
- <u>wave size</u> depends on:
  1. wind **speed**
  2. **duration** of storm
  3. **fetch** (distance over which the storm winds blow)
- a **fully developed sea** = maximum wave size for the storm conditions

Swell
- beyond the "sea"
- <u>little loss of energy</u>
- more regular in shape
- **dispersion** of wave trains (longer L waves = faster)
- **interference patterns** from swell generated in other storms

Shallow Water
- orbital motion of wave energy encounters <u>friction</u> with seafloor
- <u>waves slow down</u> and "pile up":
  $\downarrow$ wavelength (L)
  $\uparrow$ wave height (H)
  $\downarrow$ celerity (L/T)
- waves "break": (H/L = 1/7)
- incoming waves set-up **longshore currents** parallel to the shore

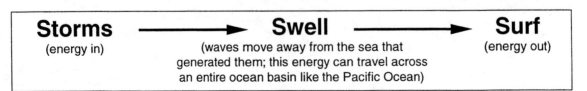

Storms ——→ **Swell** ——→ **Surf**
(energy in)

(waves move away from the sea that generated them; this energy can travel across an entire ocean basin like the Pacific Ocean)

(energy out)

Deep Water Waves (depth>L/2)

wave period (T) = the time it takes two successive crests to pass a fixed point

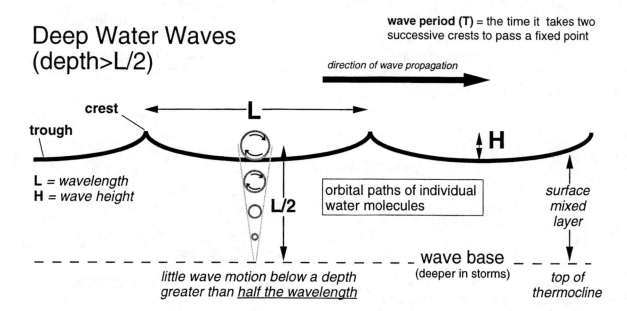

direction of wave propagation

crest

trough

L

L = *wavelength*
H = *wave height*

H

orbital paths of individual water molecules

surface mixed layer

L/2

little wave motion below a depth greater than <u>half the wavelength</u>

wave base
(deeper in storms)

top of thermocline

in deep water (D>L/2), wave velocity, or **celerity (C)**, is controlled by wave period **(T)** and wavelength **(L): C = L/T**, also, **C = 1.25L$^{1/2}$ or C = 1.56T**

Intermediate (depth<L/2) and Shallow Water Waves (depth<L/20)

*waves break in the **surf zone** when the ratio of H/L is 1/7*

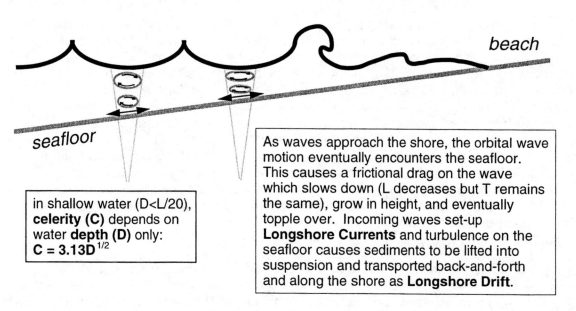

beach

seafloor

in shallow water (D<L/20), **celerity (C)** depends on water **depth (D)** only:
C = 3.13D$^{1/2}$

As waves approach the shore, the orbital wave motion eventually encounters the seafloor. This causes a frictional drag on the wave which slows down (L decreases but T remains the same), grow in height, and eventually topple over. Incoming waves set-up **Longshore Currents** and turbulence on the seafloor causes sediments to be lifted into suspension and transported back-and-forth and along the shore as **Longshore Drift**.

Tides: Earth's Attraction to the Moon & Sun

The **gravitational attraction of the Moon and the Sun** produce ocean tides, which are long wavelength, shallow water, waves (see p. 142-143). However, tide-generating forces are very complex. The gravitational attraction between two bodies is proportional to their masses, and is inversely proportional to the square of the distance (radius^2) between them. In other words, gravitational attraction drops off quickly with increasing distance. Despite its smaller mass, the Moon has about twice the tide-generating force as the Sun because it is so close to Earth.

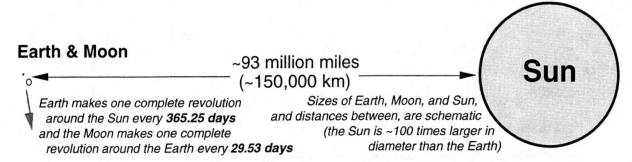

Earth & Moon

~93 million miles
(~150,000 km)

Sun

*Earth makes one complete revolution around the Sun every **365.25 days** and the Moon makes one complete revolution around the Earth every **29.53 days***

Sizes of Earth, Moon, and Sun, and distances between, are schematic (the Sun is ~100 times larger in diameter than the Earth)

The Earth and Moon revolve about a common center of mass. This produces **two tidal bulges** on the surface of the ocean: one bulge is drawn toward the moon and a second (same size) develops on the opposite side of the Earth resulting from the centrifugal force of the rotating Earth-Moon system. Like two skaters facing each other, holding hands, and rotating in a circle, the force necessary to grasp hands and stay together is equal to the centrifugal force that is trying to send them flying backwards onto their backsides.

When the gravitational forces of the Moon and Sun are in the same direction (new moon and full moon), the tidal range is maximum (higher high tides and lower low tides). These are called **spring tides** and they occur every-other week. When the Moon and Sun are in quadrate positions (first and third quarter of the moon), then tidal range is minimal producing what are called **neap tides**. Neap tides alternate every-other week with spring tides.

The **equilibrium model** of tides predicts two high tides and two low tides a day as the Earth passes under the two tidal bulges in its daily rotation. However, the presence of continents and irregularities of coastlines and continental margins complicates the daily movement of the two tidal bulges. The **dynamical model** of tides accounts for the observed patterns of tidal movement. In some ocean basins, the tidal bulge (wave) moves progressively down the ocean basin. This scenario occurs in the South Atlantic and is referred to as a **progressive tide** because the time of high tide, for example, gradually migrates northward along the coasts of South America and Africa with each tidal cycle. In other ocean basins, such as the North Atlantic, the tidal bulge is confined to a particular region. In this case, the wave rotates around a node in the central part of the basin where there is no tidal influence. This node is called the **amphidromic point** and the pattern of tidal bulge movement is called a **rotary standing wave** because of the progression of each high tide around the coastline of the North Atlantic in a counter-clockwise pattern (the rotation is clockwise in the Southern Hemisphere).

Some coasts, like those around the Gulf of Mexico, experience a once daily tidal cycle, called a **diurnal tide**. A diurnal tide has a tidal period of 24 hours, 50 minutes. The east coast of the U.S. has a twice-daily tide, called a **semidiurnal tide**, with a tidal period of 12 hours, 25 minutes. The most common type is a **mixed tide** with either one or two tidal cycles a day as exemplified by the west coast of the U.S.

144

all diagrams below represent polar views looking down on the Earth's axis of rotation (North Pole) and the plane of the Moon's orbit around Earth; the Sun is to the right

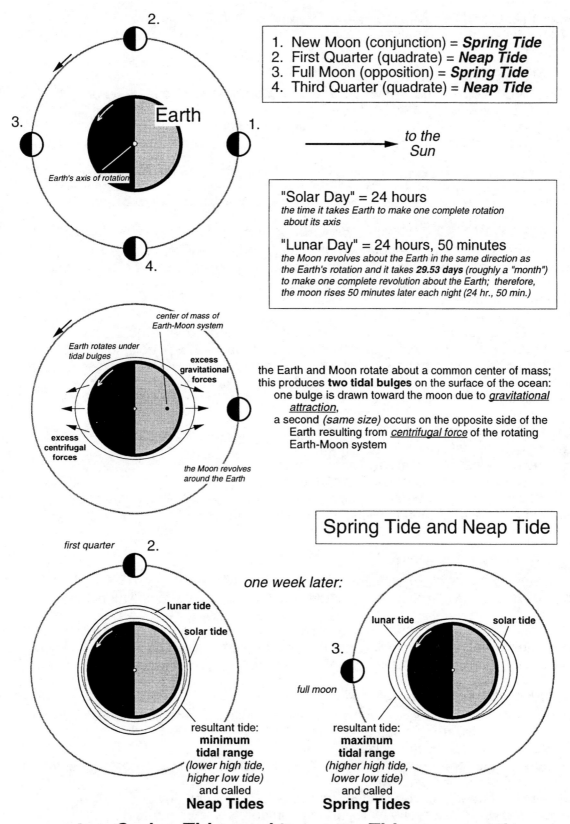

2.

Earth

3.

Earth's axis of rotation

1.

4.

1. New Moon (conjunction) = *Spring Tide*
2. First Quarter (quadrate) = *Neap Tide*
3. Full Moon (opposition) = *Spring Tide*
4. Third Quarter (quadrate) = *Neap Tide*

to the Sun

"Solar Day" = 24 hours
the time it takes Earth to make one complete rotation about its axis

"Lunar Day" = 24 hours, 50 minutes
*the Moon revolves about the Earth in the same direction as the Earth's rotation and it takes **29.53 days** (roughly a "month") to make one complete revolution about the Earth; therefore, the moon rises 50 minutes later each night (24 hr., 50 min.)*

center of mass of Earth-Moon system

Earth rotates under tidal bulges

excess gravitational forces

excess centrifugal forces

the Moon revolves around the Earth

the Earth and Moon rotate about a common center of mass; this produces **two tidal bulges** on the surface of the ocean:
   one bulge is drawn toward the moon due to *gravitational attraction*,
   a second *(same size)* occurs on the opposite side of the Earth resulting from *centrifugal force* of the rotating Earth-Moon system

Spring Tide and Neap Tide

first quarter

2.

one week later:

lunar tide

solar tide

resultant tide:
minimum tidal range
(lower high tide, higher low tide)
and called
Neap Tides

lunar tide   solar tide

3.

full moon

resultant tide:
maximum tidal range
(higher high tide, lower low tide)
and called
Spring Tides

two Spring Tides and two neap Tides per month

5 Kingdoms of Life

adapted from "A Teacher's Guide to Accompany the 5 Kingdoms of Life Poster",
written by Louise Armstrong and Lynn Margulis

Kingdom **MONERA*** *(bacteria)*
- most are TINY CELLS (~1 micron) which lack a membrane-bounded nucleus and contain 'naked' DNA (= **PROKARYOTIC CELLS**)
- **HETEROTROPHIC** bacteria (consumers and decomposers)
- **AUTOTROPHIC** bacteria (photosynthetic cyanobacteria and chemosynthetic bacteria)

*some biologists prefer to split this group into two Kingdoms, **EUBACTERIA** ("true bacteria") and **ARCHAEA** ("old bacteria"); ARCHAEA are distinguished by unique aspects of their biochemistry, and many forms tolerate extreme conditions (very high temperatures, high salinity, acidic and/or anoxic waters)*

Kingdom **PROTOCTISTA** *(algae, protists, slime molds)*
- cells which have a membrane-bounded NUCLEUS, and usually contain internal organelles such as mitochondria, plastids, and golgi bodies (= **EUKARYOTIC CELLS**; advanced cells evolved via bacterial symbiosis)
- not bacteria, not fungi, not plant, not animal
- most are larger than bacteria
- diverse in structure and feeding (**AUTOTROPHIC** and **HETEROTROPHIC** forms)
- many are **UNICELLULAR** (single-celled organisms called PROTISTS) including all PHYTOPLANKTON in the ocean and other microscopic protozoans ("microzooplankton")
- **MULTICELLULAR** forms include green, brown, and red ALGAE ("seaweeds")

Kingdom **FUNGI** *(yeasts, molds, mushrooms)*
- all have NUCLEATED CELLS
- develop from SPORES which are resistant to drying
- mostly TERRESTRIAL (living on land in moist air)
- require food in the form of organic compounds (HETEROTROPHIC like animals) but digest food outside rather than inside their bodies by releasing enzymes onto their food and decomposing it

Kingdom **PLANTAE** *(plants)*
- develop from an EMBRYO surrounded by tissue of female parent
- all are **MULTICELLULAR** and each nucleated cell is covered by a cell wall composed of cellulose
- most conduct **PHOTOSYNTHESIS**: produce oxygen and use green pigment chlorophyll to make their own food by reducing carbon dioxide (**AUTOTROPHIC**)
- marine varieties include salt marsh grasses, turtle and eel grasses, mangrove trees

Kingdom **ANIMALIA** *(animals)*
- develop from a type of embryo called a BLASTULA (multicellular hollow sphere) formed when an egg is fertilized by a sperm
- all are **MULTICELLULAR** with nucleated cells
- require food in the form of organic compounds (**HETEROTROPHIC**), other organisms or the remains of other organisms
- includes invertebrates and vertebrates; sponges to squid, and tiny zooplankton crustaceans to huge cetaceans (whales) and humans (see facing page)

Kingdom **ANIMALIA**
Major groups (PHYLA) of marine invertebrate
and vertebrate animals

Invertebrates

PORIFERA *(sponges)*
- primitive suspension feeders with digestive cavity; intertidal to abyss

CNIDARIA *(corals, jellyfish, sea anemones, comb jellies)*
- radial symmetry, possess stinging cells; corals build $CaCO_3$ skeleton

PLATYHELMINTHES *(flat worms)*
- bilateral symmetry, primitive central nervous system

NEMOTODA *(round worms)*
- flow-through digestive system

ANNELIDA *(segmented worms)*
- includes tube worms and feather duster worms

MOLLUSCA *(chitons, snails, clams, oysters, mussels, squid, octopi)*
- bilateral symmetry; flow-through digestive tract; well-developed nervous system
- most have internal or external shell of calcium carbonate ($CaCO_3$)

ARTHROPODA *(crabs, shrimp, lobsters, barnacles, krill, copepods)*
- appendages with joints; organism grows by molting
- exoskeleton of tough chitin (N-rich carbohydrate), some strengthened with $CaCO_3$

ECHINODERMATA *(sea stars, sea urchins, sea cucumbers)*
- pentameral symmetry; water-vascular system; tube feet for locomotion and feeding
- some with external shell of calcium carbonate ($CaCO_3$)

CHORDATA *(tunicates, salps)*
- possess stiffened nodocord

Vertebrates

CHORDATA
  VERTEBRATA (Subphylum)
- cartilaginous fishes *(sharks, skates, rays)*
- bony fishes *(tuna, halibut, sea horse, eel)*
- amphibians *(no marine representatives)*
- reptiles *(turtles, crocodiles, sea snakes, marine iguanas)*
- birds *(gulls, albatrosses, petrels, penguins)*
- mammals *(whales, seals, otters, manatees)*

Life in the Ocean: the Effects of Salinity & Temperature

Seawater poses a special problem for many marine organisms because of a difference in **ionic concentration** (salinity) between the body fluids of an organism and its salt water environment. This becomes particularly problematic for organisms that migrate seasonally to waters of the world ocean with different salinities, or for **anadromous fish** such as salmon, which mature in the ocean but spawn in the same fresh water streams where they were born. Cell walls are **semipermeable**, which means that they allow some molecules to pass through and screen others out. Molecules can move into and out of cells by a process called **diffusion**. Diffusion is the passive movement of molecules from high concentration to low concentration; no energy is exerted (watch what happens when you put a tea bag in a cup of hot water). The diffusion of water molecules into or out of a cell is called **osmosis**. If there is a salinity gradient between the inside and outside of the cell, an **osmotic pressure** will cause water molecules to move from high concentration of water (low salinity) to low concentration of water (higher salinity).

The osmotic pressure experienced by many salt water fish and marine mammals is caused by cells and tissue having an ionic concentration less than the seawater they live in. This condition is called **hypotonic** and these organisms have evolved strategies to overcome the loss of water (dehydration) from their cells (see facing page). The opposite problem is faced by fresh water fish because their cells have a higher ionic concentration than their environment. This condition is referred to as **hypertonic**. Sharks, rays, and many marine invertebrates experience little osmotic pressure because their body fluids have an ionic concentration, which is very close to that of the seawater they live in. This condition is called **isotonic**.

Many marine organisms are sensitive to relatively small changes in the physical characteristics of their environment. **Salinity** can significantly impact the distribution of some marine organisms because of osmotic pressures. **Stenohaline** organisms can tolerate only a narrow range of salinity. For example, many organisms would not be able to tolerate the high salinities (>40°/$_{oo}$) of some subtropical lagoons, or the reduced salinities (<30°/$_{oo}$) of coastal waters or estuaries. **Euryhaline** organisms can tolerate a wider range of salinities. For example, many organisms that live along the coasts of the mid-latitudes must be able to tolerate daily and seasonal swings in salinity due to the tidal fluctuations, evaporation, precipitation, and river runoff from the land.

Temperature also has a significant impact on the distribution of marine organisms. For example, rates of diffusion, osmosis, and metabolism are temperature-dependent. The higher the temperature the higher the rate of molecular movement into or out of cells, and the higher the rate of biological activity including growth rates, motility, and life span. Temperature also controls the concentration of dissolved gasses in water including CO_2 for photosynthesis and O_2 for animal respiration (see Primary Productivity); the higher the temperature, the less dissolved gas water can hold. **Stenothermal** organisms can tolerate only a narrow range of temperatures and **eurythermal** organisms can tolerate a wider range. Coral reefs and mid-latitude intertidal communities are two examples of shallow water communities representing the end-members of stenohaline-stenothermal and euryhaline-eurythermal organisms, respectively.

Temperature also affects the **viscosity** of water. Viscosity relates to fluids' internal resistance to flow. Cold maple syrup has a relatively high viscosity (it resists flow) compared to warmed maple syrup, which not only goes great with blueberry pancakes, but has a significantly lower viscosity (it flows easily). The viscosity of water has an important impact on marine organisms, particularly floating organisms such as plankton and swimming organisms such as fish, marine mammals, and reptiles.

Diffusion

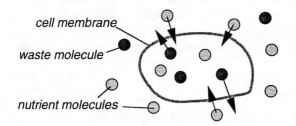

cell membrane

waste molecule

nutrient molecules

diffusion: <u>high</u> concentration → <u>low</u> concentration

∴ **nutrients in,
wastes out**

Osmosis

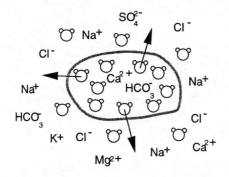

SO_4^{2-} Cl^-

Na^+

Cl^-

Ca^{2+} Na^+

HCO_3^-

Na^+

HCO_3^- Cl^-

K^+ Cl^-

Mg^{2+} Na^+ Ca^{2+}

salinity outside > salinity inside
*(water is more concentrated on the
inside of the cell relative to "salts")*

∴ **water moves out***

***hypotonic cells**
as in marine fish

Organisms are adapted to the conditions of the environment in which they live:	*Tropical Coral Reefs*	*Mid-Latitude Intertidal Zone*
salinity	~34°/oo-37°/oo	~0°/oo-40°/oo
temperature	~18°-29°C (~65°- 84°F)	~0°-25°C (~32°-77°F)
organism tolerances	**stenohaline, stenothermal** *(narrow range)*	**euryhaline, eurythermal** *(broad range)*

The Distribution of Life in the Ocean

The **pelagic environment** refers to the <u>water column</u> (from the surface to the bottom); the **benthic environment** refers to the <u>seafloor</u> (from a salt marsh or beach to the deepest trench). There are many more benthic species of animals (~98% of all animal species) than pelagic species (~2%) because of the greater variety of habitats available for exploitation and specialization on the seabed compared with the water column (see p. 160-167).

Plankton include all organisms that <u>drift with the ocean currents</u>: *"passive floaters"*
- **phytoplankton** - microscopic, single-celled, photosynthetic algae (*diatoms, dinoflagellates, coccolithophorids*)
- **zooplankton** - includes some animals (*copepods, krill, ctenophores, jellies, salps, arrow worms*) and microscopic, single-celled protists (*flagellates, ciliates, foraminifera, radiolaria*)
- **bacterioplankton** - includes some photosynthetic forms (*cyanobacteria*)
- **meroplankton** - larval stage of some benthonic and nektonic animals (spend early part of life in the plankton)

Nekton are all organisms capable of <u>moving independent of ocean currents</u>: *"swimmers"*
- squid, chambered nautilus, fish (*pelagic fish such as herring, anchovy, macheral, tuna, marlin, shark, rays, tropical reef fish, flying fish, salmon, eel, and "ground fish" such as cod and haddock*), marine mammals (*seals, manatees, toothed whales such as dolphin and sperm whale, baleen whales such as the Humpback, Right, Fin, Gray, and Minke whales*), marine reptiles (*sea turtles, sea crocodiles*)

Benthon are all organisms that live <u>on the seafloor</u> (**epifauna**) or <u>buried within sediments</u> (**infauna**): *"bottom-dwellers"*
- most marine invertebrates (*clams, mussels, oysters, scallops, limpets, snails, barnacles, lobsters, shrimp, crabs, sea urchins, starfish, brittle stars, sea cucumbers, corals, anemones, sponges, worms*)
- attached plants (*seagrasses*) and benthic algae (*kelp and other seaweeds*)
- some fish (*flounder, sole*)

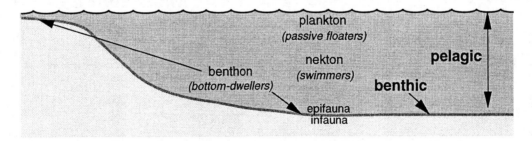

Seawater and living tissue have nearly the same density because most plants, animals, bacteria, and protists, consist mostly of water and dissolved solids ("salts"). The similarities in density between seawater and living matter confer several important advantages for marine organisms. For example, some organisms float or sink only slowly because they have nearly neutral buoyancy, and heavy skeletons are not needed for support. Seawater also contains abundant dissolved gases and mineral nutrients.

Despite the fact that organisms are composed mostly of water, most organisms will have a tendancy to sink in seawater. The base of most food chains in the ocean, like those on land, is dependent on photosynthesis, which is dependent on sunlight (see p. 152-153). Therefore, adaptations for flotation or maintaining position in the water column, particularly in sunlit surface waters where photosynthesis is possible, are crucial facts of life for many marine organisms, from the tiny phytoplankton to the herbivores to the biggest predators:

The temperature, and hence the **viscosity**, of water has an important impact on marine organisms, particularly floating organisms such as plankton and swimming organisms such as fish, marine mammals, and reptiles. Cold water is more viscous than warm water, therefore it is easier to float in cold water, but harder to swim. For the plankton, the problem of maintaining position in the water column is more difficult in the sunlit surface waters of low to mid-latitudes because warm waters are less dense and less viscous than colder waters.

MAINTAINING POSITION IN THE WATER COLUMN

Adaptations by plankton (passive floaters):
1. large surface area (A) to volume (V) ratio (= large surface area, like a parachute):
- **tiny size** (the smaller the size, the greater the A/V ratio)
- **appendages** or **spines** (greater surface area, ∴ greater A/V ratio)
- **flat shapes** (= greater surface area)
- **form chains** (= greater surface area)

2. produce **oil droplets** within protoplasm (= greater buoyancy)

Adaptations by nekton (swimmers):
1. active swimming in fast predators (expend energy to overcome gravity)
2. swim bladders to regulate buoyancy (i.e., to rise or sink in the water column) in not-so-active swimmers and benthic fishes
3. gas containers to regulate buoyancy in cephalopod molluscs (*chambered nautilus, squid, cuttlefish*)

Primary Productivity in the Ocean

Consider the nature of food chains and food webs on land: plants are at the base of most terrestrial food webs. Plants are **photosynthetic autotrophs**; they reduce *inorganic carbon* to produce *organic molecules* utilizing the energy of the sun to drive the biochemical reactions of **photosynthesis**. Oxygen (O_2) is released as a by-product of photosynthesis. Plants depend on supplies of **carbon dioxide** (CO_2), **water**, **nutrients** (especially <u>nitrates</u> and <u>phosphates</u>, but also trace elements such as iron, as well as vitamins), and **sunlight**. Photosynthesis can be summarized by the following reaction:

$$6CO_2 + 6H_2O + \text{inorganic nutrients} + \text{solar energy} \rightarrow C_6H_{12}O_6 + 6O_2$$

carbon dioxide   water   glucose (simple sugar)   oxygen

the opposite of this reaction is **respiration** *(what animals do exclusively)*

Chemosynthetic autotrophs, such as the bacteria at hydrothermal vents and hydrocarbon seeps on the seafloor, produce organic molecules by utilizing chemical reactions, rather than sunlight, as the energy source (see p. 164-165).

The production of organic matter by photosynthetic and chemosynthetic autotrophs is called **primary productivity**. All animals are dependent on primary productivity, either by feeding directly on autotrophs in the case of **herbivores** ("grazers"), or by feeding indirectly through predation by **carnivores**, or by scavenging dead organisms (see p. 158-159). The **organic matter** produced by autotrophs is <u>stored chemical energy</u> in the form of carbohydrates, proteins, and fats *(for example, you might eat a candy bar for a quick shot of energy, or load-up on "carbos", which are easily broken-down to sugars that provide energy for a busy day of work or play).*

The vast majority of primary productivity in the oceans is the result of **photosynthesis**, and much of this is by **microscopic single-celled algae** collectively called **phytoplankton** (autotrophic protists). Other important **primary producers** include **cyanobacteria** (photosynthetic bacteria) in near-surface waters of the world ocean, as well as **multicellular benthic algae** (seaweeds such as kelp) and **vascular plants** (such as salt marsh, eel, and turtle grasses, and mangrove trees) in coastal waters.

Primary productivity by <u>photosynthesis requires two essential ingredients</u>, **solar energy** and **inorganic nutrients**. If both are not readily available, productivity will be limited. The **photic zone** corresponds with the depth to which photosynthesis is possible; 20 m (~66 ft.) or so in coastal waters and rarely exceeding 200 m (~660 ft.) in the tropical ocean (see p. 160). The **euphotic zone** is the upper part of the photic zone that receives enough light to support a net gain in photosynthetic production (the **oxygen compensation depth** defines the base of the euphotic zone; this is where autotroph consumption of O_2 by respiration exceeds production of organic matter and O_2 by photosynthesis - see Exercise #24).

The *availability of solar energy* varies across **latitude** and it varies by **season**, particularly in the mid- to high latitudes because of the differences in the **angle of solar incidence**. The magnitude of surface water heating (hence the degree of thermocline development) and the depth of the photic zone depend on the angle of solar incidence; the higher the sun is in the sky, the warmer the surface waters become and the deeper sunlight will penetrate into the ocean.

The *availability of nutrients* in the surface waters of the open ocean is significantly influenced by the density structure of the upper water column. For example, in the tropics, nutrients are concentrated in the deep waters below the euphotic zone and below the thermocline (see facing page). The presence or absence of a **seasonal thermocline** in the temperate belt is important because it may limit the availability of nutrients for part of the year (see p. 156-157). You can think of the thermocline as a "<u>density doorway</u>"; when closed (year-round or seasonal thermocline) nutrient availability to the euphotic zone is limited, when open (weak or absent thermocline) nutrient diffusion to the euphotic zone is not limited. The weak thermal and density stratification of polar waters does not prohibit the upward diffusion of nutrients at any time of the year. However, the low sun angle and large seasonal changes in number of daylight hours controls the timing of productivity in the high latitudes (see p. 156-157).

Angle of solar incidence varies
with **latitude** and **season**

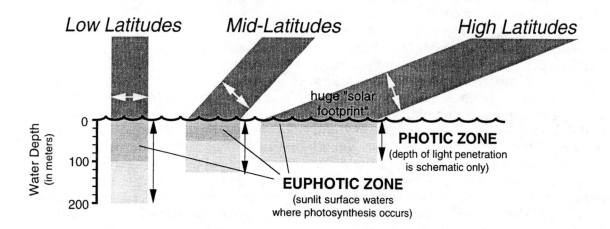

Low Latitudes   Mid-Latitudes   High Latitudes

huge "solar footprint"

Water Depth (in meters)

0

100

200

PHOTIC ZONE
(depth of light penetration
is schematic only)

EUPHOTIC ZONE
(sunlit surface waters
where photosynthesis occurs)

The **thermocline** can be an effective
**barrier to the diffusion and
vertical advection of nutrients**
between nutrient-rich deep waters
and the sunlit surface waters

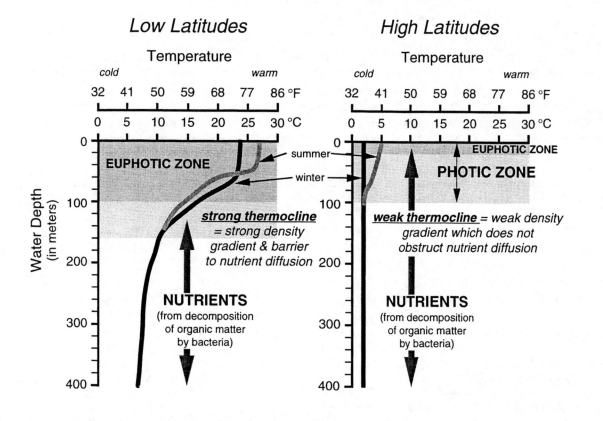

Low Latitudes

Temperature

cold   warm
32 41 50 59 68 77 86 °F

0   5   10 15 20 25 30 °C

0

100

200

300

400

Water Depth (in meters)

EUPHOTIC ZONE

summer

winter

strong thermocline
= strong density
gradient & barrier
to nutrient diffusion

NUTRIENTS
(from decomposition
of organic matter
by bacteria)

High Latitudes

Temperature

cold   warm
32 41 50 59 68 77 86 °F

0   5   10 15 20 25 30 °C

0

100

200

300

400

EUPHOTIC ZONE

PHOTIC ZONE

weak thermocline = weak density
gradient which does not
obstruct nutrient diffusion

NUTRIENTS
(from decomposition
of organic matter
by bacteria)

Nutrients & the Marine Carbon Cycle

The **nutrients** required by phytoplankton, algae, and marine plants to produce organic matter are the same as those needed by terrestrial plants, namely, **nitrates** (NO_3), **phosphates** (PO_4), **trace elements** such as iron (Fe), and **vitamins**. The primary source of dissolved nutrients is from the weathering of soils and rocks on the land. Organisms, in general, are opportunistic; if there are nutrients available and there is adequate solar radiation, then primary productivity will be high (see p. 152). Therefore, the **greatest biomass** (greatest concentration of organisms) is along the continental margins, especially **coastal waters**, close to the source of nutrients (rivers). Surface waters are typically stripped of their nutrients by the autotrophs, although there are some important exceptions. As a consequence, limited concentrations of dissolved nutrients make it out into the open ocean surface waters beyond the continental margins. These waters support little biomass because of insufficient nutrient supply. In effect, the **subtropical gyres** of the world ocean are **biological deserts**. Areas of **oceanic divergence** and **upwelling** at the edges of the gyres provide elevated concentrations of essential nutrients along these relatively narrow bands of surface waters (see p. 134).

Once organic matter is produced, **heterotrophic bacteria** living in the water column and at the seafloor break it down again, thereby recycling the nutrients back to their usable inorganic form. This process of releasing nutrients back to the water column is called **bacterial degradation** (see p. 158). However, once in the deep waters, dissolved nutrients do not easily diffuse upward into the sunlit surface waters to be utilized by the phytoplankton because of the stratified nature of the water column in the low to mid-latitudes. The nutrients become "trapped" in the deep waters below the pycnocline (thermocline), except where **upwelling** occurs. In this way, deep waters are a **sink**, or a storehouse, for nutrients. Upwelling provides a mechanism to deliver nutrients, the raw ingredients of primary productivity, back to the sunlit surface waters (see p. 134).

In addition to bacterial degradation, organic matter from the sunlit surface waters will begin to oxidize and decompose as it rains down through the water column. **Oxidation** causes oxygen depletion at depth and the formation of **oxygen minimum zones**, especially along continental margins where surface productivity is high. The oxygen minimum zone (<4 ml/l, milliliters of dissolved oxygen per liter of seawater) is generally restricted to the upper part of the water column (~300-1300 m) because deep waters below the pycnocline are very cold and oxygen-rich (see p. 122-123, 138-139).

The **biological pump** describes a complex suite of biological and biochemical processes that result in the export of **atmospheric CO_2** to the deep ocean as **organic carbon** and biologically precipitated **calcium carbonate** ($CaCO_3$). Through the process of **photosynthesis**, phytoplankton and cyanobacteria manufacture organic carbon from inorganic carbon dioxide (CO_2). Some phytoplankton (such as coccolithophorids), other protists (foraminifera), and many groups of invertebrate animals (including molluscs, echinoderms, and corals) manufacture carbonate shells from dissolved calcium (Ca^{2+}) and bicarbonate (HCO_3^-) ions. This dissolved CO_2 (and HCO_3^-) is derived in part from the atmosphere due to the action of wind and wave. A significant proportion of the organic matter and carbonate created in the euphotic zone is *exported out of the surface waters* as **fecal pellets** or other forms of particulate organic matter called **marine snow**.

The biological pump is an important component of the **carbon cycle**, whereby carbon in a variety of forms is utilized by organisms for a variety of purposes, and is then recycled back to its inorganic form to be utilized again. For example, some of the organic matter produced in the euphotic zone is buried in marine sediments, but much of it is eventually consumed by heterotrophic organisms or decomposed by bacteria, both of which liberate CO_2 as a by-product of respiration and decomposition. Much of the dissolved CO_2 will eventually find its way back to the surface ocean in areas of upwelling where it can interact with the atmosphere once again. Above the level of the **carbonate compensation depth** (CCD), carbonate shells are deposited on the seafloor as sediments called **calcareous ooze** (see p. 108-109). Below the CCD, the carbonate shells are dissolved, and Ca^{2+} and HCO_3^- are returned to the water column.

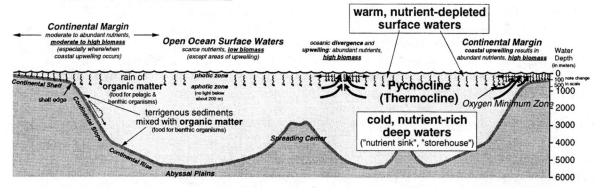

Inorganic Nutrients:
Sources, Sinks, and Recycling

warm, nutrient-depleted surface waters

Continental Margin
moderate to abundant nutrients, **moderate to high biomass** *(especially where/when coastal upwelling occurs)*

Open Ocean Surface Waters
scarce nutrients, **low biomass** *(except areas of upwelling)*

oceanic divergence and upwelling: abundant nutrients, **high biomass**

Continental Margin
coastal upwelling results in abundant nutrients, **high biomass**

Water Depth (in meters)

Continental Shelf

shelf edge

rain of **organic matter** *(food for pelagic & benthic organisms)*

photic zone
aphotic zone (no light below about 200 m)

Pycnocline (Thermocline)

Oxygen Minimum Zone

Continental Slope

terrigenous sediments mixed with **organic matter** *(food for benthic organisms)*

cold, nutrient-rich deep waters ("nutrient sink", "storehouse")

Continental Rise

Spreading Center

Abyssal Plains

0
100 note change
500 in scale
1000
2000
3000
4000
5000
6000

Marine life is concentrated in sunlit surface waters where photosynthesis occurs; however, vast areas of the surface ocean (e.g., subtropical gyres) are depleted in the nutrients required for photosynthesis due to the presence of a pycnocline (thermocline). The rain of organic matter from surface waters, as well as the influx of terrestrial organic material from the continents, provide food for pelagic and benthic organisms that live below the euphotic zone. Therefore, greater concentrations of pelagic and benthic organisms occur beneath areas of higher primary productivity and terrestrial organic carbon input compared with the water column and seafloor beneath the subtropical gyres.

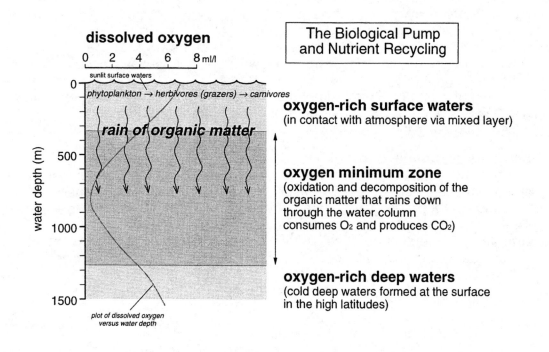

dissolved oxygen

The Biological Pump and Nutrient Recycling

0   2   4   6   8 ml/l

sunlit surface waters

phytoplankton → herbivores (grazers) → carnivores

rain of organic matter

water depth (m)

0

500

1000

1500

plot of dissolved oxygen versus water depth

oxygen-rich surface waters
(in contact with atmosphere via mixed layer)

oxygen minimum zone
(oxidation and decomposition of the organic matter that rains down through the water column consumes O_2 and produces CO_2)

oxygen-rich deep waters
(cold deep waters formed at the surface in the high latitudes)

Seasonality & Primary Productivity

The abundance of marine life tracks the distributional patterns of productivity. For example, where there is a great abundance of autotrophs (primary producers), then it follows that there will be a great abundance of animals (**high biomass**) (see p. 158-159). If the availability of nutrients or sunlight is limited, then there will be relatively few primary producers and therefore few animals (**low biomass**). **Biomass** is a term that biological oceanographers use to describe the total weight (mass) of all organisms, or of a particular group of organisms, in an environment or habitat. In general, the highest productivity (therefore the highest biomass) is along continental margins, especially in coastal waters near rivers, the principle source of ocean nutrients, or in areas of coastal upwelling or oceanic divergence (see p. 134-135).

Seasonality is a major control on the distribution of marine life along continental margins and in the open ocean. The <u>presence or absence of a strong thermocline</u> ("density doorway"), including the development of a **seasonal thermocline** in the mid-latitudes, combined with **seasonal changes in solar radiation** (angle of incidence), greatly influence primary productivity across latitude.

1. High Latitudes (Polar Belt)
* productivity is not limited by availability of *nutrients* because a very weak pycnocline (thermocline) means ready-access to nutrients ("density doorway open")
* productivity is <u>limited by the availability of *solar radiation*</u> through much of the year
* a <u>short-lived episode of productivity occurs in late spring-early summer</u> when there is more solar energy (sunlight available 24 hours a day)
* therefore, productivity is *solar-limited* for much of the year in polar waters

2. Mid-Latitudes (Temperate Belt)
* **winter:** nutrients are available ("door open") but the angle of solar incidence is too low to drive high productivity; productivity<u> is *solar-limited*</u>
* **spring:** enough nutrients and sunlight = <u>high productivity</u> (**"spring bloom"**)
* **summer:** warming of surface waters and development of strong seasonal thermocline cuts off the nutrient supply from below ("density doorway closed"); productivity<u> is *nutrient-limited*</u>
* **fall:** breakdown of seasonal thermocline as surface waters begin to cool which allows nutrients to "leak" into the photic zone ("door partially opened"), still enough light = <u>second pulse of productivity</u>

3. Low Latitudes (Tropical Belt)
* there is always enough solar radiation, but as a consequence of the warm surface waters and strong thermocline the nutrient supply is cut-off from below ("density doorway closed tight"); productivity<u> is *nutrient-limited*</u> <u>year-round</u>, except where there is <u>upwelling</u> (see p. 134-135)
* **coral reefs** are an exception because highly efficient ecosystem-level recycling of nutrients supports high productivity, despite very low concentrations of dissolved nutrients in the warm, clear waters (see p. 166).

Seasonal Changes in Productivity Across Latitude

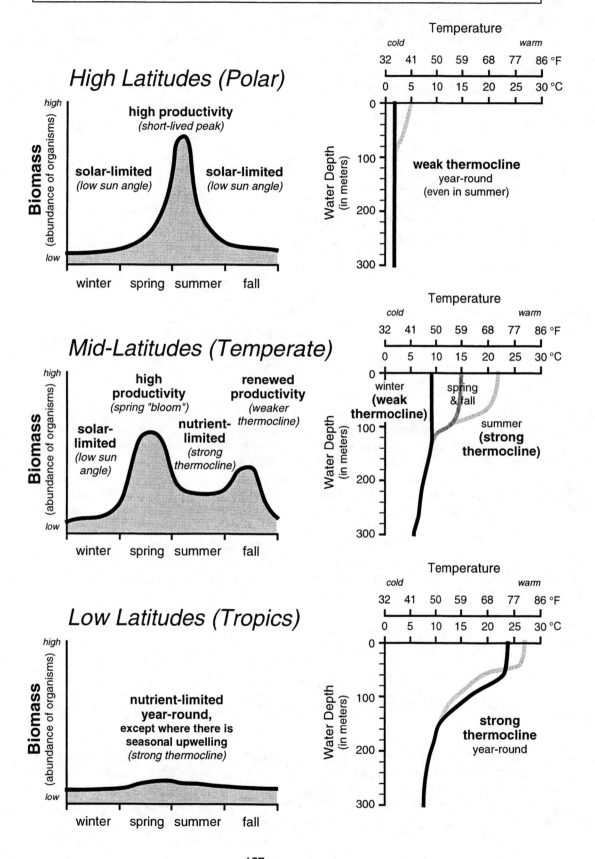

High Latitudes (Polar)

Biomass (abundance of organisms)

high — low

high productivity
(short-lived peak)

solar-limited
(low sun angle)

solar-limited
(low sun angle)

winter   spring   summer   fall

Temperature

cold ... warm

32 41 50 59 68 77 86 °F

0 5 10 15 20 25 30 °C

Water Depth (in meters)

0 — 100 — 200 — 300

weak thermocline
year-round
(even in summer)

Mid-Latitudes (Temperate)

Biomass (abundance of organisms)

high — low

high productivity
(spring "bloom")

renewed productivity
(weaker thermocline)

solar-limited
(low sun angle)

nutrient-limited
(strong thermocline)

winter   spring   summer   fall

Temperature

cold ... warm

32 41 50 59 68 77 86 °F

0 5 10 15 20 25 30 °C

Water Depth (in meters)

0 — 100 — 200 — 300

winter **(weak thermocline)**

spring & fall

summer **(strong thermocline)**

Low Latitudes (Tropics)

Biomass (abundance of organisms)

high — low

nutrient-limited year-round,
except where there is
seasonal upwelling
(strong thermocline)

winter   spring   summer   fall

Temperature

cold ... warm

32 41 50 59 68 77 86 °F

0 5 10 15 20 25 30 °C

Water Depth (in meters)

0 — 100 — 200 — 300

strong thermocline
year-round

Food Webs & the Trophic Pyramid

A **food chain** is a sequence of organisms representing a feeding hierarchy; one organism is food for the next organism in the sequence. **Predators** (the hunters) feed on live **prey** (the hunted), while **scavengers** feed on dead and decaying animal or plant remains. In reality, feeding relationships are much more complex than a simple food chain because many organisms consume multiple types of prey. A **food web** describes the intricate nature of multiple interacting food chains in a community and the **flow of energy** from the **producers** (autotrophs) to the **consumers** to the **decomposers**.

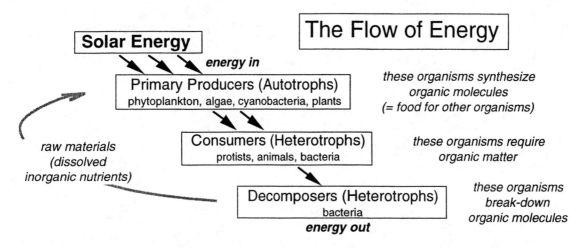

The Flow of Energy

Solar Energy → energy in

Primary Producers (Autotrophs)
phytoplankton, algae, cyanobacteria, plants

these organisms synthesize organic molecules
(= food for other organisms)

Consumers (Heterotrophs)
protists, animals, bacteria

these organisms require organic matter

raw materials (dissolved inorganic nutrients)

Decomposers (Heterotrophs)
bacteria
energy out

these organisms break-down organic molecules

Using a simple generalized example of a food chain we can distinguish four trophic levels: 1) **primary producers** (phytoplankton, **autotrophs**), 2) **herbivores** (grazers, **heterotrophs**), 3) **carnivores** (predators, also heterotrophs) or **scavengers**, and 4) **top carnivore**. **Trophic levels** represent successive stages of nourishment (energy consumption). Only ~6-15% of energy consumed at each trophic level goes into creating biomass (i.e., *stored chemical energy*: carbohydrates, proteins, and fats) available for the next trophic level to consume and benefit from. This is referred to as **Transfer Efficiency**. The remainder of the energy is lost to other metabolic functions such as respiration, feeding, digestion, locomotion, and reproduction. If we assume that this transfer efficiency is approximately 10%, then it follows that the biomass of each trophic level differs by a factor of 10 (ten times as much phytoplankton biomass as grazer biomass). Therefore, a much larger biomass must be available to support (feed) the next trophic level, hence the **Trophic Pyramid**.

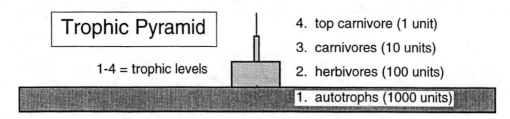

Trophic Pyramid

1-4 = trophic levels

4. top carnivore (1 unit)

3. carnivores (10 units)

2. herbivores (100 units)

1. autotrophs (1000 units)

Ecology is the study of organisms and their interactions with their environment and each other. An **ecosystem** includes all the organisms in a particular environment; from the primary producers to the top carnivores, as well as the diverse communities of bacteria, algae, protists, and fungi that play vital roles in food webs and the recycling of carbon and nutrients.

Food Chains/Food Webs

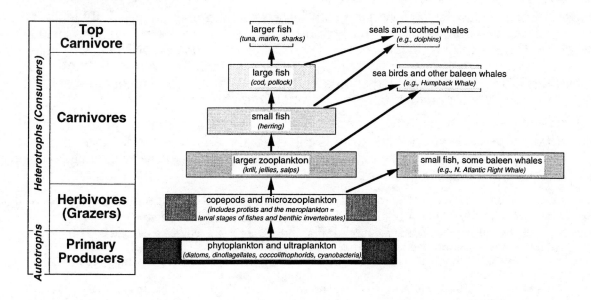

Trophic Pyramid

several examples of trophic pyramids

if assume a **transfer efficiency of ~10%**, then the biomass of each trophic level differs by a factor of 10; therefore, ***a very large biomass is necessary to support the next trophic level***

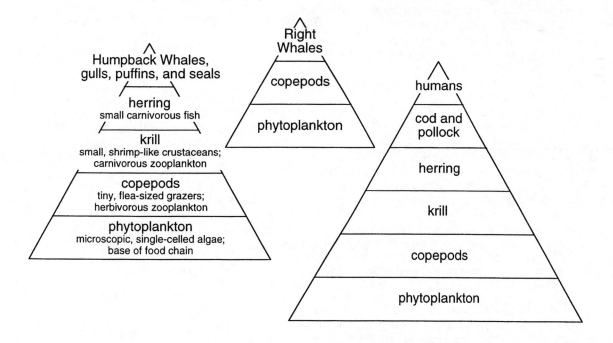

Coastal Ecosystems: Highly Productive But Vulnerable

Coastal habitats are especially rich in life because of the abundance of nutrients that stimulates vigorous productivity. The primary producers in coastal ecosystems include representatives of three kingdoms of life: 1) autotrophic bacteria (MONERA), 2) single-celled protists and multi-cellular algae (PROTOCTISTA), and 3) true plants (PLANTAE). **Detritus** is also an important food source in coastal waters; it consists of dead and decaying algae and/or grasses covered with heterotrophic bacteria.

The **littoral zone** is the area between low tide and high tide; it is also called the **intertidal zone**. The **supratidal zone** is the area just above high tide; this area is affected daily by salt water spray along rocky coasts, or occasionally by exceptionally high tides or by storm surge along other parts of the coast.

There are a wide variety of environments and habitats for organisms along the coasts of the world. The following is a brief list of major coastal environments:
- **estuaries** - where freshwater meets seawater (mouths of rivers, bays, fjords; see p. 174)
- **river deltas** - accumulations of sediment at the mouth of a river
- **salt marshes** - vegetated tidal mud flats found along shores protected from surf (see p. 174)
- **tidal mud flats** - intertidal areas choked with mud
- **beaches & dunes** - high energy shorelines where sand is always on the move (see p. 172-173)
- **barrier islands & spits** - low-lying ribbons of sand created by longshore drift (see p. 172-173)
- **lagoons** - shallow body of water separated from the open ocean by a barrier island/spit, or reef
- **mangrove swamps** - tropical intertidal waters vegetated by mangrove trees
- **coral reefs** - rigid structures of calcium carbonate ($CaCO_3$) built by colonial corals and calcareous algae that support biologically diverse communities of organisms in the tropical photic zone (p. 166)

Coastal habitats are some of the richest environments known on Earth. **High biological productivity** supports large communities. Two types of "oases" can be distinguished by the level of nutrients that supports the communities.
- abundant nutrients: "fertile" oases (***salt marshes & estuaries; mangrove swamps***)
- scarce nutrients: "fragile" oases (***coral reefs***)

These environments are important because they:
- support many **coastal & oceanic food webs**
- provide **nurseries** for many species of fish
- supply **food** for humans
- provide productive habitats along **flyways** for migratory birds
- stimulate **ecotourism**

Loss of these precious resources through environmental degradation and development will cause untold damage to commercial fisheries, and to marine ecosystems in general.

BENTHIC LIFE ALONG THE SHORE
A. Rocky Shores (including *tide pools*)
- high energy (pounding of the surf)
- many of the <u>animals live attached</u> to rocks and seaweed (= **epifauna**)
- these organisms have very specific tolerances to atmospheric exposure
- this is the most easily recognizable vertical biozonation in the marine environment, for example:
  - *periwinkle snails:* supratidal zone (spray zone, above high tide)
  - *barnacles, limpets:* high tide zone (dry most of day)
  - *mussels:* mid-tide zone (half-day dry/half day wet)
  - *attached algae, anemones, sea stars:* low tide zone (mostly wet)

B. Sediment-Covered Shores
- high energy (*beaches*) and low energy (*salt marshes, tidal mud flats*)
- many of the <u>animals live buried</u> within the sediment (= **infauna**)
- intertidal zonation not as easily recognizable
- examples of infaunal organisms:
  - *clams, echinoderms (like sand dollars), crabs, worms*
- examples of epifaunal organisms:
  - *snails (gastropods), mussels, salt marsh grasses*

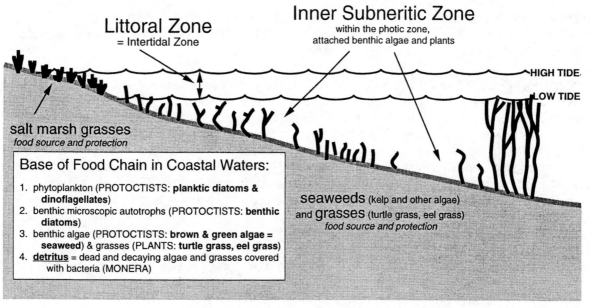

Coastal Habitats

Littoral Zone
= Intertidal Zone

Inner Subneritic Zone
within the photic zone,
attached benthic algae and plants

HIGH TIDE

LOW TIDE

salt marsh grasses
food source and protection

seaweeds (kelp and other algae)
and **grasses** (turtle grass, eel grass)
food source and protection

Base of Food Chain in Coastal Waters:

1. phytoplankton (PROTOCTISTS: **planktic diatoms & dinoflagellates**)
2. benthic microscopic autotrophs (PROTOCTISTS: **benthic diatoms**)
3. benthic algae (PROTOCTISTS: **brown & green algae = seaweed**) & grasses (PLANTS: **turtle grass, eel grass**)
4. <u>detritus</u> = dead and decaying algae and grasses covered with bacteria (MONERA)

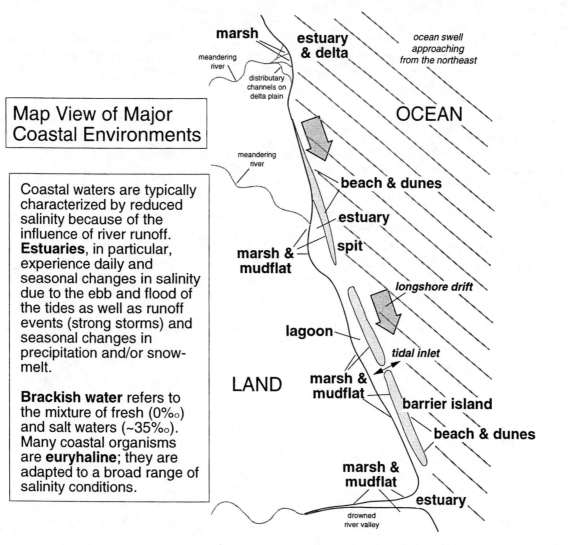

Map View of Major Coastal Environments

Coastal waters are typically characterized by reduced salinity because of the influence of river runoff. **Estuaries**, in particular, experience daily and seasonal changes in salinity due to the ebb and flood of the tides as well as runoff events (strong storms) and seasonal changes in precipitation and/or snow-melt.

Brackish water refers to the mixture of fresh (0‰) and salt waters (~35‰). Many coastal organisms are **euryhaline**; they are adapted to a broad range of salinity conditions.

marsh

estuary & delta

ocean swell approaching from the northeast

meandering river

distributary channels on delta plain

OCEAN

meandering river

beach & dunes

estuary

marsh & mudflat

spit

longshore drift

lagoon

tidal inlet

LAND

marsh & mudflat

barrier island

beach & dunes

marsh & mudflat

estuary

drowned river valley

Continental Shelf Ecosystems

The **sublittoral zone** is the area of the shelf below low tide. The terms **neritic province** and **subneritic province** refer to the water column over the continental shelf and the seafloor of the continental shelf, respectively.

The **continental shelf** is a dynamic oceanographic environment because of the relatively shallow depths and the seasonal changes in primary productivity, water mass characteristics, and storminess. Both pelagic and benthic organisms are influenced by these daily or seasonal changes, including the effects of tidal mixing and storm waves. Under fair weather conditions, the influence of wave motion extends to 5-15 m (~16-49 ft.). This is called **fair weather wave base** (see p. 142-143). **Storm wave base** typically reaches to depths of 20-30 m (~66-99 ft.), but can extend to >200 m (657 ft.) in severe storms. The seafloor is unaffected by waves below the storm wave base.

Benthic plant life is restricted to the inner part of the shelf because the depth of the **euphotic zone** is typically much shallower than further offshore. For example, river runoff and sediment turbidity (sediments in suspension due to wave and tidal energy) clouds the water thereby reducing clarity and turning it a brownish color. An abundance of dissolved river-borne nutrients stimulates photosynthesis and the resulting phytoplankton biomass further limits light penetration in coastal waters. These waters may appear greenish in color due to the abundance of chlorophyll pigment in the phytoplankton. Tidal mixing and storms reintroduce nutrients back into the euphotic zone thereby helping to sustain high productivity on the shelf. The euphotic zone may be less than 20 m (66 ft.) in turbid coastal waters and as deep as ~150 m (492 ft.) in the "blue" waters of the tropical open ocean.

BENTHIC LIFE OFFSHORE
A. Rocky Bottom
- attached benthic algae (e.g., *kelp*) on the inner shelf where sunlight can reach the seafloor
- epifaunal animals dominate, examples:
  lobsters, echinoids (sea urchins), oyster beds, snails
B. Sediment-Covered Bottom
- marine grasses occur on the inner shelf where sunlight can reach the seafloor
- infaunal and epifaunal animals, examples:
  clams, scallops, worms, sea stars, horseshoe crabs, echinoids (sand dollars), snails, shrimp

PELAGIC LIFE ON THE SHELF
- base of food chain: *diatoms, dinoflagellates*
- grazers: *protists, copepod*
- carnivores: *squid, fish, jellyfish, arrow worms*
- seasonal changes in productivity result in seasonal changes in pelagic and ground fish stocks:
  pelagic fishes: *anchovy, mackerel*; ground fish: *cod, haddock, hake, flounder, sole*
- where productivity is high, seasonal migratory visitors may include species of toothed (*dolphin*) and baleen whales (*Humpback, Gray, Right, Fin, Blue*)

Neritic Habitats

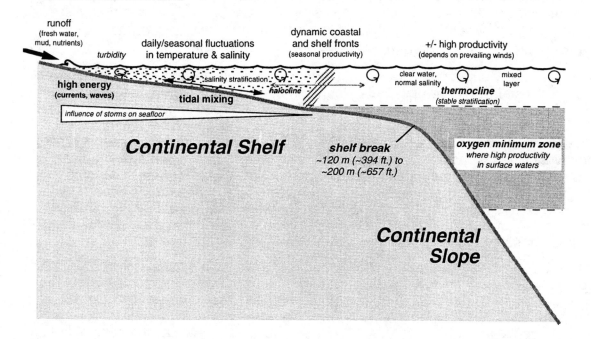

runoff
(fresh water, mud, nutrients)

turbidity

daily/seasonal fluctuations in temperature & salinity

dynamic coastal and shelf fronts
(seasonal productivity)

+/- high productivity
(depends on prevailing winds)

high energy
(currents, waves)

salinity stratification

clear water, normal salinity

mixed layer

tidal mixing

halocline

thermocline
(stable stratification)

influence of storms on seafloor

Continental Shelf

shelf break
~120 m (~394 ft.) to
~200 m (~657 ft.)

oxygen minimum zone
where high productivity
in surface waters

Continental Slope

Water mass **fronts**, like the passage of the warm and cold fronts that control our weather, are created on the shelf where waters of different origins and characteristics meet. For example, a turbid (muddy) coastal water mass with its reduced salinity will be in contact with a clear open ocean water mass with normal salinity. They may also have different temperature characteristics. If there are density differences between the two water masses, the denser of the two will sink below the other to produce an inclined front (see above). If the two water masses are moving in opposite directions, they will slide past each other with relatively little mixing along a vertical front. This is the case where the warm **Gulf Stream** waters move north along the continental margin of the eastern U.S. and then northeastward into the North Atlantic, while the cold, nutrient-rich **Slope Water** moves southward between the Gulf Stream and the continental shelf. Occasionally, **warm core rings** spin off the Gulf Stream and interact with the Slope Water and coastal water masses on the shelf (see p. 132). Productivity can be elevated along fronts where mixing introduces nutrients to the euphotic zone. Shelf fronts are dynamic; the position of a front can shift hourly, weekly, or seasonally.

In areas characterized by **coastal upwelling** and high productivity over the outer shelf and upper slope, an **oxygen minimum zone** may impinge on the shelf. These conditions yield organic-rich muddy sediments. In cases of low terrigenous sedimentation rates, the nutrient-rich waters will precipitate **phosphorite deposits** on the seafloor (see p. 108-109).

Open Ocean ("Blue Water") Ecosystems

The greatest concentration of primary producers is in sunlit waters. Therefore, many animals live close to their food source in the upper water column. This feeding relationship can be generalized by the **"grazing food web"** in photic zone communities:

phytoplankton → *zooplankton* → *nekton* → *bigger nekton*

Primary productivity supported by an external supply of nutrients, such as river-borne nutrients in coastal waters or upwelled nutrients in open ocean divergences, is called **new production**. The nutrient-poor surface waters of the open ocean, the so-called "biological deserts" of the subtropical gyres, are characterized by **regenerated production** which refers to productivity sustained by an internal supply of recycled nutrients via bacterial degradation within the photic zone. The **microbial loop** is the food web based on diverse communities of bacteria, including autotrophic cyanobacteria, and single-celled protist consumers that characterize such **"blue water" ecosystems** (no turbidity, scarce nutrients, low phytoplankton biomass = clear "blue" water). These ecosystems cannot sustain large communities of animal zooplankton and fish due to the scarce nutrient supply.

Animals that live primarily below the photic zone (i.e., below the depths of living phytoplankton) depend on **fecal material**, **marine snow**, and other forms of **detritus** (dead and decaying aggregates of organic matter) raining down from above. Detritus is also transported from shallower waters in **turbidity currents** which roll downslope as fast-moving, dense mixtures of sediment and water. Detritus serves as the base of the **"detritus food web"** in aphotic pelagic communities:

detritus → *nekton* → *bigger nekton*

and in many benthic communities:

detritus → *scavengers and deposit feeders* → *nekton*

BENTHIC LIFE IN THE DEEP-SEA
- seafloor beyond the continental shelf (= *suboceanic province*)
- dark, no light, *no photosynthetic autotrophs*
- very low temperatures (rarely >4°C)
- diversity (number of species) is high, including microorganisms such as protists (*despite scarcity of food!!*)
- echinoderms (*sea stars, sea urchins, sea cucumbers*) more common than molluscs (*clams, snails*)
- limiting factor on deep-sea biomass = availability of food (food supply may be highly seasonal)
- food of deep-sea animals and protists: bacteria, other microorganisms, other animals, rain of organic matter from surface waters, including detritus, fecal pellets, and marine snow (aggregates of organic particles and bacteria)

A. Low Biomass Communities:
Abyssal Plains
- scarce food but very high diversity of animals, protists, and bacteria; the environment is stable over long (geologic) time scales and the organisms have specialized by partitioning the limited resources available to them.

B. High Biomass Communities:
1. Under areas of high surface water productivity (*many continental margins, oceanic divergence*)
   - abundant food (rain of organic matter through the water column)
2. "Chemosynthetic Oases": isolated, ephemeral communities (short-lived)
   - *chemosynthetic bacteria* (= autotrophs) are the base of the food chain:
      a. **hydrothermal vent communities** (at spreading centers)
      b. hydrocarbon seeps (gas hydrates: methane CH_4)
      c. large carcasses (e.g., whale falls)

Deep-Sea Benthic and Pelagic Habitats

much of the open ocean is a **biological desert** *(e.g., surface waters beyond the continental margins and the floor of the deep-sea)*; **low productivity** in surface waters means low animal biomass and limited rain of organic matter for benthic organisms

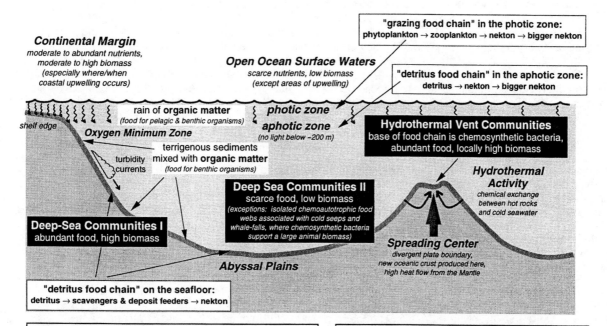

Continental Margin
moderate to abundant nutrients, moderate to high biomass (especially where/when coastal upwelling occurs)

Open Ocean Surface Waters
scarce nutrients, low biomass (except areas of upwelling)

"grazing food chain" in the photic zone:
phytoplankton → zooplankton → nekton → bigger nekton

"detritus food chain" in the aphotic zone:
detritus → nekton → bigger nekton

shelf edge

rain of **organic matter**
(food for pelagic & benthic organisms)

photic zone
aphotic zone
(no light below ~200 m)

Oxygen Minimum Zone

turbidity currents

terrigenous sediments
mixed with **organic matter**
(food for benthic organisms)

Hydrothermal Vent Communities
base of food chain is chemosynthetic bacteria, abundant food, locally high biomass

Hydrothermal Activity
chemical exchange between hot rocks and cold seawater

Deep Sea Communities II
scarce food, low biomass
(exceptions: isolated chemoautotrophic food webs associated with cold seeps and whale-falls, where chemosynthetic bacteria support a large animal biomass)

Deep-Sea Communities I
abundant food, high biomass

Abyssal Plains

Spreading Center
divergent plate boundary, new oceanic crust produced here, high heat flow from the Mantle

"detritus food chain" on the seafloor:
detritus → scavengers & deposit feeders → nekton

Photoautotrophic Food Webs:
photosynthetic phytoplankton and cyanobacteria, limited only by the availability of inorganic nutrients and solar radiation, support surface-dwelling organisms, as well as pelagic and benthic organisms via rain of organic matter through the water column

Chemoautotrophic Food Webs:
hydrothermal vents, cold seeps, and whale-falls represent isolated, ephemeral communities in the deep-sea; chemosynthetic bacteria support other organisms in the absence of sunlight; organisms grow rapidly and produce abundant larvae

---

PELAGIC LIFE IN BLUE WATER
- "biological deserts" of the subtropical gyres (low nutrient) vs. high productivity of continental margins and oceanic divergences (moderate to high nutrient)
- downward flux of organic matter (**biological pump**) controls the biomass of organisms below the photic zone

A. Photic Zone (enough solar radiation to power photosynthesis)
- base of food chain in low nutrient waters: *coccolithophorids, cyanobacteria*; in higher nutrient waters: *diatoms, dinoflagellates*
- diverse, but generally low biomass communities of organisms
    zooplankton: *copepods, euphausiid shrimp, jellyfish, tunicates, arrow worms*
    fish and other animals include: *squid, marlin, mahi mahi (dolphin fish), flying fish, sharks, dolphin, and migratory animals including many species of whale, sea turtles, tuna, salmon, eel*

B. Dysphotic & Aphotic Zones (very low to no solar radiation; *no photosynthesis*)
- fish with small, large mouths, expandable bodies, some with photophores producing **bioluminescent** spots or bioluminescent "lures"
    lantern fish, hatchet fish, anglerfish, gulper

Coral Reef Ecosystems: the Rainforests of the Sea

Coral reefs are wave resistant, rigid structures of calcium carbonate ($CaCO_3$) built by colonial corals and calcareous algae that support biologically diverse communities of organisms in the tropical photic zone. Coral reefs develop in warm (>18°C or 65°F), clear waters of normal to slightly elevated salinity. Water clarity is due to the little terrigenous sediment input (mud) from river runoff, and scarce dissolved nutrients to support a phytoplankton biomass.

Coral reefs, like tropical rainforests, are among the most biologically diverse ecosystems on the planet, and like the rainforests, coral reefs are highly productive despite the limited availability of nutrients. Both of these ecosystems are highly efficient at recycling nutrients and capitalizing on **symbiotic relationships** at all levels of the intricate food webs, from the bacteria and other microbes to the invertebrate and vertebrate animals. **Symbiosis** is a term to describe the intimate co-existence of two different organisms, or the dependence of one organism on another. Symbioses are common in the coral reef ecosystem. For example, microscopic symbiotic algae (dinoflagellate protists called **zooxanthellae**) exist in the tissue and cells of corals and foraminifera (protists), respectively. The algae and their host organisms both benefit. The algae receive protection from predators and the metabolic waste products of the hosts serve as a nutrient supply in a nutrient-poor world. The corals and foraminifera benefit because the zooxanthellae provide an internal source of oxygen for respiration as well as a food supply provided by photosynthesis. The symbionts also assist in the precipitation of their calcium carbonate colonies and shells. Other examples of symbiotic relationships in the reef include the clown fish and the anemone, cleaner fish and moray eel, and remora fish and shark.

Under environmental stresses, linked in part with <u>elevated sea surface temperatures</u> (**SSTs**), corals and foraminifera will expel their zooxanthellae. In the colonial corals this process is called **coral bleaching** because the normally greenish or brownish color of the living coral animals turns a ghostly white. If the SSTs return to normal within a matter of months, the corals and foraminifera will be reinoculated with dinoflagellate symbionts and recover fully. But if elevated SSTs persist for several years, massive coral die-off can occur and the health of the reef ecosystem will be compromised. **Global warming** is but one threat to coral reef ecosystems; sediments associated with clear-cutting of tropical forests, chemical pollutants, and nutrient loading are also threatening reef ecosystems around the world.

In general, a coral reef is characterized by a high-energy **reef core** (or reef terrace) composed of the frame-building corals, as well as calcareous algae, which encrust and bind the reef structure. At its seaward edge, the reef grows up to sea level where the stony structure provides turbulence and vigorous mixing. A low energy **back reef** and shallow **lagoon** may contain small **patch reefs** (see below). Variable conditions of water depth, turbulence, light level, and nature of the substrate across the reef profile create a great variety of "ecospaces" (**niches**) for organisms to occupy.

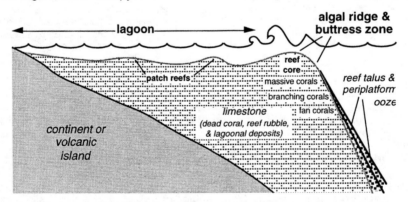

BENTHIC LIFE IN THE CORAL REEF
corals, sponges, calcareous algae, foraminifera (protists), anemones (& clownfish), bryozoans, crown-of-thorns seastars, sea urchins, nudibrachs, barnacles, crabs, sea urchins, featherduster worms, sea fans, brittle stars, soft corals, tunicates, giant clams, cowry snails, shrimp

PELAGIC LIFE IN THE CORAL REEF
moray eel, angelfish, scorpion fish, bumphead parrot fish, butterfly fish, puffer fish, trigger fish, grouper, sharks, rays, sea nettles,

Indo-Pacific coral reefs are more diverse than **Atlantic reefs** and calcareous algae are a minor frame-builder of Atlantic reefs. Atlantic reefs lack many of the invertebrate species of the Indo-Pacific, including numerous coral species, the giant clam *Tridacna*, and the diverse communities of molluscs and crustacea.

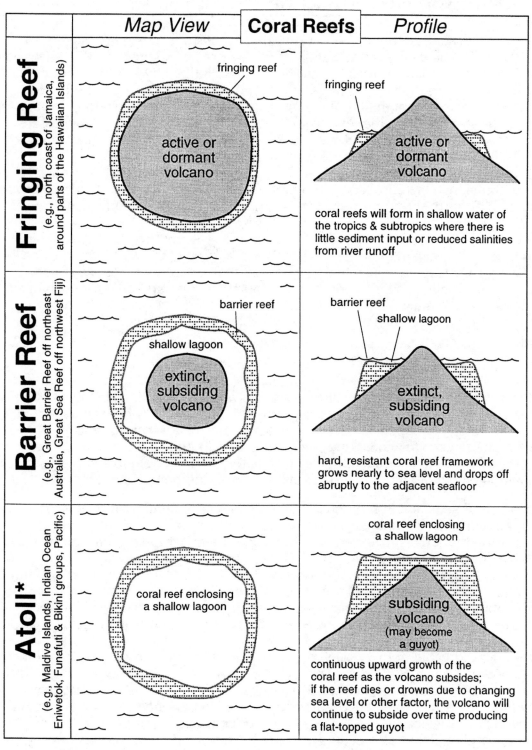

Map View	Coral Reefs	Profile

Fringing Reef (e.g., north coast of Jamaica, around parts of the Hawaiian Islands)

fringing reef

active or dormant volcano

fringing reef

active or dormant volcano

coral reefs will form in shallow water of the tropics & subtropics where there is little sediment input or reduced salinities from river runoff

Barrier Reef (e.g., Great Barrier Reef off northeast Australia, Great Sea Reef off northwest Fiji)

barrier reef

shallow lagoon

extinct, subsiding volcano

barrier reef

shallow lagoon

extinct, subsiding volcano

hard, resistant coral reef framework grows nearly to sea level and drops off abruptly to the adjacent seafloor

Atoll* (e.g., Maldive Islands, Indian Ocean Eniwetok, Funafuti & Bikini groups, Pacific)

coral reef enclosing a shallow lagoon

coral reef enclosing a shallow lagoon

subsiding volcano (may become a guyot)

continuous upward growth of the coral reef as the volcano subsides; if the reef dies or drowns due to changing sea level or other factor, the volcano will continue to subside over time producing a flat-topped guyot

*The explanation for the origin of atolls, as depicted above as the gradual sinking of a tropical volcanic island, was first proposed by **Charles Darwin** during his travels aboard the *Beagle* (1831-1836).

Sediment in the Coastal Zone

Sediments of the coastal zone include a wide variety of unconsolidated particles of inorganic or organic (biogenic) origin. These particles are transported to the present-day coast by wind, water, or ice. A rocky coastline is characterized by **boulders, cobbles, and pebbles**, which are derived from the erosion of sea cliffs or exposed bedrock. Along the New England coast, much of the rocky coastline owes its origin to the debris left by retreating ice sheets during the last episode of continental glaciation. Beaches are typically composed of rounded **sand-size** rock fragments, mineral grains such as quartz and feldspar (the two most common rock-forming minerals in the Earth's crust), or biogenic particles such as whole or broken shells. These grains are rounded smooth by high-energy abrasion in the surf zone. The **mud** found in estuaries, marshes, and bays is composed of **clay and silt-sized** particles together with variable amounts of **organic detritus**.

The **grain size** of a sedimentary deposit reflects the <u>energy</u> of the wind, water, or ice that transported the particles, as well as the environment that allowed the particles to be deposited. For example, sand and pebbles accumulate in the surf zone, although much of this material may be moved again by storm waves and longshore currents, but turbulence is too great for the finer grained particles (silt, clay) to accumulate.

boulders → cobbles → pebbles → sand → silt → clay

high energy   →   *low energy*

Much of the sand and mud of the coastal zone is delivered to the ocean by rivers. Some sediment is derived from the erosion of coastal areas, such as rocky headlands, sea cliffs, and ancient sedimentary deposits. **Erosion** of the coast is caused by pounding surf, tides, severe storms, strong winds, and rising sea level. In addition, the wind delivers sand and finer material to the coast but plays a much more important role in the redistribution of sand in a natural give-and-take between the beach and dunes.

Transportation of sediment in the coastal zone occurs by a number of processes including wind, currents, tides, and gravity. Sand is transported parallel to shore via **longshore currents**. The movement of sand down the coast is called **longshore drift** or **longshore transport** (see p. 170-171). Once sand is deeper than about 10 m (~30 ft.), it is lost from the longshore transport system. The sand is then transported by storm waves and gravity to deeper parts of the continental shelf and beyond (see p. 163). The finer grained mud (silt and clay) is more easily held in suspension and is therefore typically transported as sediment plumes seaward of the high-energy surf zone.

Deposition of sand in high energy environments like beaches, barrier islands, and spits is mostly temporary; much of the sand is shuffled between the beach, dunes, and offshore sand bars in a natural seasonal cycle that maintains the profile of the shoreline provided the supply of sand has not been severed or disrupted by man-made structures such as dams on rivers, or by jetties, groins, break-waters, or seawalls along the coast. Deposition of sand and mud also occurs in deltas where sedimentation rates are high, and in quieter water settings such as salt marshes, estuaries, and bays. Mud also accumulates offshore on the continental shelf, beyond the reach of the high energy surf zone.

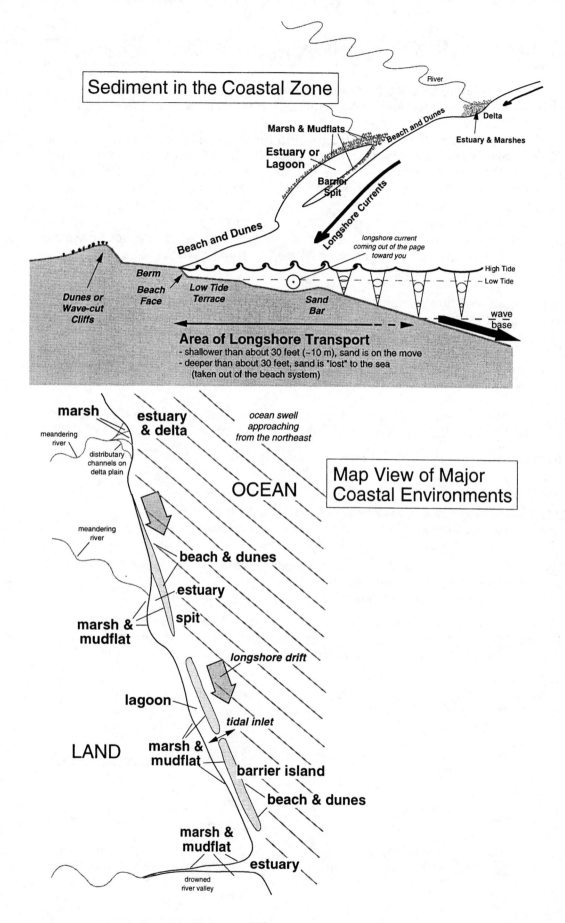

Sediment in the Coastal Zone

River

Delta

Marsh & Mudflats

Beach and Dunes

Estuary or Lagoon

Estuary & Marshes

Barrier Spit

Longshore Currents

Beach and Dunes

longshore current coming out of the page toward you

High Tide

Low Tide

Dunes or Wave-cut Cliffs

Berm

Beach Face

Low Tide Terrace

Sand Bar

wave base

Area of Longshore Transport
- shallower than about 30 feet (~10 m), sand is on the move
- deeper than about 30 feet, sand is "lost" to the sea
  (taken out of the beach system)

marsh

estuary & delta

ocean swell approaching from the northeast

meandering river

distributary channels on delta plain

OCEAN

Map View of Major Coastal Environments

meandering river

beach & dunes

estuary

marsh & mudflat

spit

longshore drift

lagoon

tidal inlet

LAND

marsh & mudflat

barrier island

beach & dunes

marsh & mudflat

estuary

drowned river valley

169

Longshore Drift & Coastal Dynamics: A Summary

The mechanical energy of waves is transferred to the ocean floor when wave motion comes in contact with the seafloor in intermediate and shallow water (depth <L/2). Friction causes the waves to slow down and sediment is put into motion. Wave energy in coastal waters sets up **longshore currents** and the resultant **longshore drift** (also called **longshore transport**) of sediment (see p. 142-143). Longshore drift occurs in the **surf zone**, the area between the shore and the breakers, and can be thought of as a "river of sand" moving down the coast. This can be seen with each breaking wave as water, sand, and an unattended beach ball move up the beach face with the **swash** (up-rush of water) and then back down again with the outgoing water. However, if you watch the beach ball's movement during successive waves you will see that it not only moves up and down the beach face, but it also moves along the shore in the direction of the longshore drift (see facing page).

Incoming ocean swell rarely approaches the shore straight on. Because of this, the shoreward part of each individual wave comes in contact with the seafloor before the seaward part of the wave which is still in deep water (water depth >L/2). Waves bend (**refract**) whenever part of the wave is in intermediate or shallow water while part of the wave is still in deep water (see p. 176-177). Therefore, the waves will bend as they approach the shore.

Below is a summary of **coastal dynamics**, particularly dealing with the ever-changing conditions along the beach, presented in the following pages (p. 170-173, 176-183).

1. **coastal erosion and river discharge supply the sediment** (mostly sand) necessary to establish and maintain beaches, dunes, and barrier islands
   - *these dynamic coastal systems protect the mainland from the ocean's energy*

2. **unconsolidated sand absorbs wave energy** in the surf zone
   - *the free movement of sand down the coast, and between the beach, dunes, and sand bars, is a natural buffer against storms and the energy of the ocean*

3. **longshore currents**, driven by incoming swell or storms, distribute the sediment along the shore (longshore drift/longshore transport)(see facing page)
   - *deeper than about 30 ft. (~10 m) the sand is lost from the beach system*

4. **seasonal changes in wave energy** along the coast result in net erosion of the beach during the winter and deposition of sand on the beach during the summer
   - *produces minimal changes in the year-to-year beach profile*

5. **mobility of sand**, and its continuous supply down the coast, is necessary for a healthy shoreline to maintain itself (despite rising sea level or severe storms)
   - *coastal environments retain their profile as they "roll" landward with rising sea level*

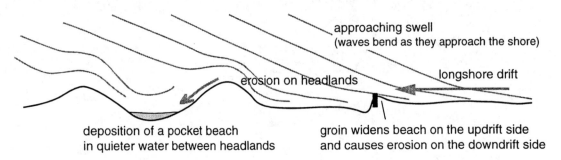

approaching swell
(waves bend as they approach the shore)

longshore drift

erosion on headlands

deposition of a pocket beach
in quieter water between headlands

groin widens beach on the updrift side
and causes erosion on the downdrift side

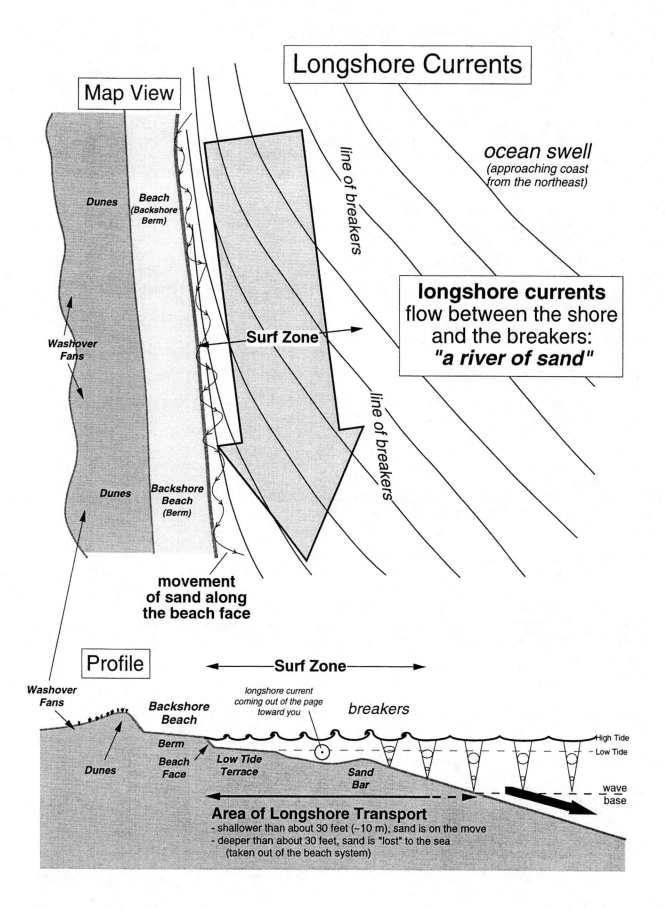

Longshore Currents

Map View

Dunes

Beach (Backshore Berm)

Washover Fans

Dunes

Backshore Beach (Berm)

Surf Zone

line of breakers

line of breakers

ocean swell
(approaching coast
from the northeast)

longshore currents
flow between the shore
and the breakers:
"a river of sand"

**movement
of sand along
the beach face**

Profile

◄─── **Surf Zone** ───►

*Washover
Fans*

**Backshore
Beach**

*longshore current
coming out of the page
toward you*

breakers

Dunes

Berm

**Beach
Face**

**Low Tide
Terrace**

**Sand
Bar**

High Tide
Low Tide

wave
base

Area of Longshore Transport
- shallower than about 30 feet (~10 m), sand is on the move
- deeper than about 30 feet, sand is "lost" to the sea
  (taken out of the beach system)

Beaches, Barrier Islands, and Spits

Beaches and **sand dunes** are accumulations of unconsolidated sand or gravel along high-energy shores. Sand is always on the move by waves, tides, longshore currents, and the wind. The free movement of sand absorbs ocean energy. Beaches and dunes are dynamic environments; in other words, they are always changing. These depositional features may look the same year after year, but considerable movement and recycling of sand occurs during the course of a year. Seasonal changes in the beach profile actually help to protect the coast from erosion and inundation by the sea (see facing page).

Barrier islands and **spits** are low-lying ribbons of sand created by longshore drift. They develop along coastlines with a wide continental shelf and an ample sand supply from river runoff or from erosion of coastal deposits. Barrier islands and spits are flood-prone and easily eroded but they serve as a natural buffer zone by protecting the mainland from wave erosion during storms. With rising sea level, these features migrate ("roll") landward as storms overwash the barriers and deposit sand on their landward sides (see p. 180-181).

Coastal zone terminology related to beach environments (see facing page):

COASTLINE - the highest elevation on the continent affected by storm waves.

Dunes - linear ridges of unconsolidated wind-blown sand that accumulate above or beyond the reach of most storm waves (landward edge of the beach). Very large storms, typically those associated with **storm surge**, can over-run the dunes to create **Overwash Fans**. These are fan-shaped accumulations of sand on the landward side of the dunes.

SHORE - the area from the low tide shoreline to the coastline. The **Backshore** is the flat area from the high tide shoreline to the coastline (=supratidal zone). The **Foreshore** is the area from the low tide shoreline to the high tide shoreline (=intertidal zone).

Beach - accumulation of unconsolidated materials along the shore including the **Berm** and foreshore **Low Tide Terrace**. The **Beach Face** (or Beach Scarp) is the sharply inclined surface between the berm and the low tide terrace that is cut by waves during high tide. Typically a beach is composed of sand-sized mineral grains (e.g., quartz and feldspar), rock fragments, or broken-up shells, but in places a beach is composed of pebble or cobble-sized rocks.

NEARSHORE - the area between the low tide shoreline and where the breakers form; also called the "surf zone".

Sand Bars - linear ridges of unconsolidated sand in the nearshore zone; they are typically dynamic, shifting location and changing shape daily or seasonally.

OFFSHORE - the area of the continental shelf seaward of the breakers.

Neritic Sediment - sand and mud, typically becoming less sandy across the continental shelf (see p. 163).

Beach and Barrier Island Profiles

Beach Profile

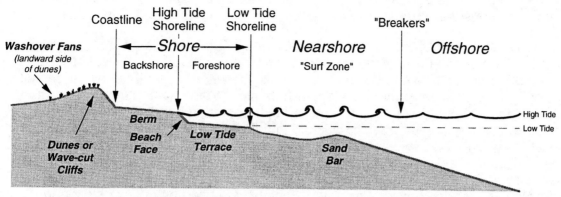

Washover Fans *(landward side of dunes)*

Coastline

High Tide Shoreline

Low Tide Shoreline

"Breakers"

Shore

Nearshore "Surf Zone"

Offshore

Backshore | Foreshore

Dunes or Wave-cut Cliffs

Berm

Beach Face

Low Tide Terrace

Sand Bar

High Tide

Low Tide

Barrier Island Profile

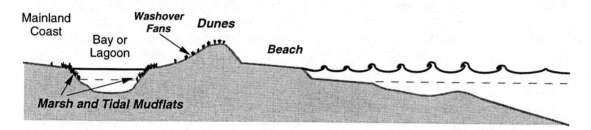

Mainland Coast

Bay or Lagoon

Washover Fans

Dunes

Beach

Marsh and Tidal Mudflats

Seasonal Cycle of the Beach

Summer Beach Profile

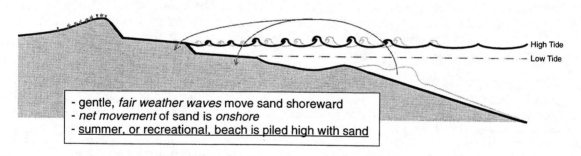

High Tide

Low Tide

- gentle, *fair weather waves* move sand shoreward
- *net movement* of sand is *onshore*
- <u>summer, or recreational, beach is piled high with sand</u>

Winter Beach Profile

sand bars "trip" the storm waves so that the waves break further offshore

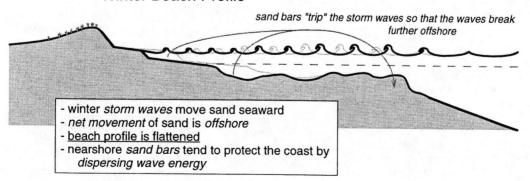

- winter *storm waves* move sand seaward
- *net movement* of sand is *offshore*
- <u>beach profile is flattened</u>
- nearshore *sand bars* tend to protect the coast by *dispersing wave energy*

Deltas, Estuaries, and Salt Marshes: Fertile Oases

River deltas are accumulations of sediment at the mouth of a river. They form where a river transports more sediment than can be carried away by longshore currents. The flood plain and the marshes that border the distributary river channel(s) are highly productive.

Estuaries are highly productive areas of the coast where the fresh water of a river meets the salt water of the ocean (e.g., mouths of rivers, drowned river valleys, bays, fjords). The salinity of the **brackish waters** (mixture of fresh and salt water) in an estuary varies daily and seasonally, as well as from top to bottom or from side to side across the estuary (see p. 160-161). In general, **estuarine circulation** is characterized by low salinity water flowing seaward over saltier water being pushed into the estuary by the tides. However, the physical, chemical, and biological characteristics vary considerably from one estuary to another. Two of the most important variables are <u>river volume</u> and <u>tidal mixing</u>. A **salt-wedge estuary** is river-dominated. As a consequence of the great volume of river water and minimal impact of tidal mixing, the estuary is highly stratified by a well-developed **halocline** (sharp change in salinity between the surface and bottom waters of the estuary). The mouths of the Mississippi River on the Gulf Coast, the Hudson River in New York City, and the Columbia River in Washington state are examples of salt-wedge estuaries. By contrast, a **well-mixed estuary** is tide-dominated. During times of the year when river volume is low, the Columbia River mouth becomes a well-mixed estuary. A **partially mixed estuary** is an intermediate between the salt wedge and well-mixed types. Chesapeake Bay, San Francisco Bay, and Puget Sound are examples.

Salt marshes are vegetated tidal flats found along shores protected from the pounding surf. Marshes border estuaries and develop at the edges of bays and lagoons behind barrier islands and spits. A dense community of salt-tolerant grasses and other plants colonize the intertidal marsh surface. The marsh surface is dissected by tidal creeks, which conduct the daily flood and ebb of the tides. Nutrients and detritus are flushed into and out of the marsh with each tidal cycle. In the tropics and subtropics, **mangrove trees** dominate the intertidal vegetation instead of grasses. The plant roots in both salt marshes and mangrove swamps extend the shoreline by stabilizing and trapping sediment. In addition, these coastal wetlands act as water filters by absorbing pollutants, and as baffles by dissipating storm energy.

The abundance of nutrients and sunlight, and the daily tidal mixing stimulates **very high biological productivity** in estuaries and marshes. The amount of organic matter produced is 4-10 times greater than an equivalent size cornfield. These rich waters support many coastal and oceanic food webs by providing spawning grounds and nurseries for many species of fish. For example, >75% of commercial fish spend at part of their life in coastal waters. Coastal waters provide food, employment, tourism and recreation. Fishing contributes ~$111 billion to the nation's economy every year. Collectively, coastal industries account for ~28 million jobs and >30% of the Gross National Product is generated in coastal counties in the U.S. About 45% of all endangered or threatened species of birds and mammals rely on coastal waters.

At times, the growth of microscopic algae (protists), typically **dinoflagellates**, is so rapid, and the abundance of cells in the water so great, that the water turns a reddish color. These algal "blooms" are called **red tides** and they pose a considerable health risk to humans and marine life of coastal waters. Some species of dinoflagellates produce toxins that become ingested by filter-feeding shellfish. The toxins then become concentrated in the tissues of the shellfish in a process called **biomagnification**. Toxins can also kill fish. Fish are also affected by the density of algal cells, which can lead to suffocation. The short-lived abundance of organic matter then settles in the estuary or bay where it decomposes and consumes the dissolved oxygen further affecting fish and benthic animals. *Pfiesteria piscicida* is a particularly nasty heterotrophic fish killing dinoflagellate species in the coastal waters of the U.S. Most of the time it occurs in low abundances or lies dormant in resting cysts at the bottom of a bay or estuary waiting for optimum conditions to emerge. Red tides occur naturally due to the right combination of nutrient-levels and stratification. However, nutrient loading in coastal waters due to sewage, fertilizers, or other agricultural runoff can also stimulate red tides.

Estuaries

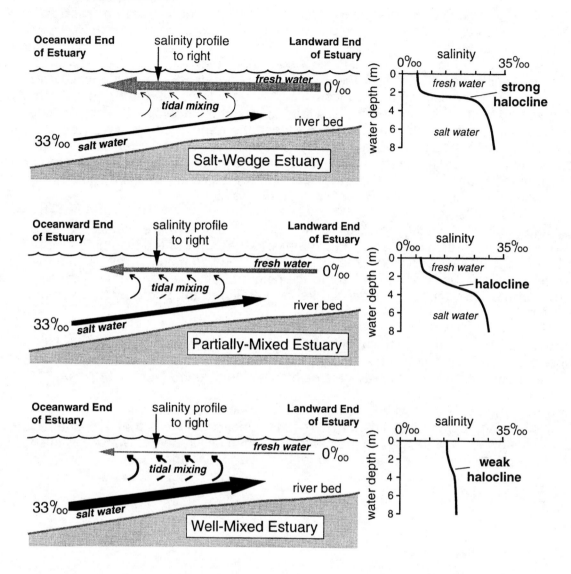

Salt Marshes

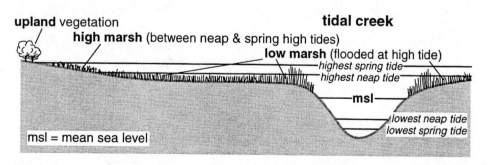

Wave Refraction: Bending of Waves

Waves will bend as they encounter shallow water. This bending is called **refraction**. As the crests of waves (swell) approach the shore, the part of the wave closest to the shore encounters shallow water before the part of the wave further offshore. As a consequence, the shoreward part of the wave begins to slow down and "pile up" while the seaward part of the wave is still in deep water (i.e., the water depth is greater than half the wavelength, depth > L/2). This bending results in a focusing of wave energy on **headlands**, parts of the coast that extend out further into the sea than other parts of the coast. This focused energy results in erosion of the headlands. Between the headlands are bays where the refraction of waves results in dispersed energy. In this way, it is easier for sediment to accumulate in the lower energy conditions of the bays (e.g., deposition of sand to form a "**pocket beach**"). Over geologic time, the coastline will be straightened by continued wave erosion on the headlands and deposition in the bays.

Along straight shorelines, you can observe the distinctive zone where breaking waves form. This so-called **surf zone** represents the area where part of the incoming swell has slowed down appreciably and where the waves have grown in height. This is also the place where the incoming waves are bent to the point of being nearly parallel with the shore and where **longshore currents** are set-up parallel to the shore by the incoming wave energy (see p. 170-171).

Waves will also **reflect** off a rigid structure like a seawall or the hull of a ship with little loss of energy. Reflected waves may then collide with on-coming waves to create larger waves and greater turbulence by way of constructive interference (see p. 182-183).

Diffraction is the bending of waves as they pass around objects such as a jetty, or as ocean swell passes between islands. The part of an individual wave that is in shallower water will be slowed down relative to the part of the wave still in deeper water. In this way, waves will bend around or between natural irregularities along the coast.

Wave Refraction

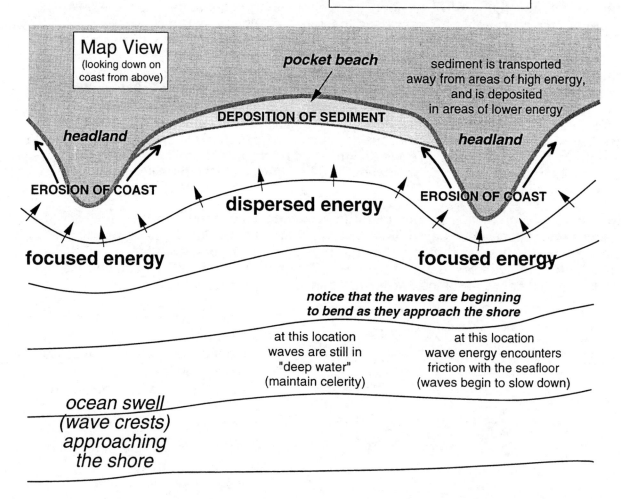

Map View
(looking down on coast from above)

pocket beach

sediment is transported away from areas of high energy, and is deposited in areas of lower energy

DEPOSITION OF SEDIMENT

headland

headland

EROSION OF COAST

EROSION OF COAST

dispersed energy

focused energy

focused energy

notice that the waves are beginning to bend as they approach the shore

at this location waves are still in "deep water" (maintain celerity)

at this location wave energy encounters friction with the seafloor (waves begin to slow down)

ocean swell (wave crests) approaching the shore

waves bend because part of a wave has entered "intermediate" water depths (D < L/2) while the remainder is in "deep" water (D > L/2)

Profile
(looking toward coast from some distance offshore)

depth < L/2

influence of waves (circular motion of water molecules)

friction between wave motion and seafloor causes waves to slow down

no wave motion below depth > L/2

seafloor

Storm Surge & Tsunamis: Coastal Hazards

Tropical cyclones (hurricanes, typhoons) are strong storms characterized by intense **low atmospheric pressure**. This causes the surface of the ocean to be drawn upwards by as much as 5-6 meters (17-20 ft.). The doming of the ocean surface and strong on-shore winds to the east of the eyewall combine to push water landward causing extensive beach erosion, coastal flooding, and property damage. This phenomenon is known as **storm surge**. The threat is made worse if "landfall" coincides with high tide. Storm surge is also associated with strong **extratropical storms** (outside the tropics, like "nor'easters" along the East Coast during the winter). During storm surge, water and waves can breach the barrier dunes causing sand to be deposited in **washover fans** on the landward side of the dunes. This is an important process in shoreline retreat as coastal environments retain their original profiles but very gradually migrate ("roll") landward (see p. 180-181).

Tsunamis represent another serious threat to some coastal communities. They have sometimes erroneously been called "tidal waves", but they have nothing to do with tides. These large waves are most often generated by **major earthquakes** associated with the **subduction** of oceanic crust (see p. 106-107). Therefore, communities around the rim of the Pacific, or islands in the Pacific, are particularly at risk.

leading edge of the overriding plate is squeezed due to drag and friction of subducting plate

gradual uplift

gradual subsidence

motion of the subducting oceanic plate

motion of the overriding continental plate

- subduction zone is locked and strain builds
- seafloor is depressed by pressure of overlying plate
- overriding plate is buckled upward due to strain

large waves

wave energy

rapid subsidence

rapid uplift

- **catastrophic release of strain** causes **major earthquake**
- seafloor rebounds rapidly as strain is removed
- rapid uplift of seafloor causes **displacement of large volume of seawater**
- wave energy disperses away from earthquake epicenter

Large waves are created as the tsunami moves across the continental shelf. Waves as large as 30 m (98 ft.) are known, but the size depends on the magnitude of the earthquake and configuration of the coastline. Tsunamis are very long wavelength waves (typically L>200 km or >124 mi) and therefore behave as shallow water waves (see p. 142-143). In the open ocean, these long, low waves travel nearly as fast as a commercial jetliner (~760 km/hr, ~480 mi/hr), where they pass without notice against the background of wind waves and swell. As they encounter the shallower waters of the continental margins, the waves slow down significantly and grow to large size. Part of the danger with tsunamis comes from the withdrawal of water near the coast, exposing the seafloor and attracting curious on-lookers. The water is drawn into the growing wave that is but minutes from rushing up and over the coast. After the first huge wave, people may think that its all over, only to be followed 5-20 minutes later by a succession of tsunami waves. The **International Tsunami Warning System** monitors earthquake activity around the world and issues warnings about potentially dangerous tsunamis.

Storm Surge

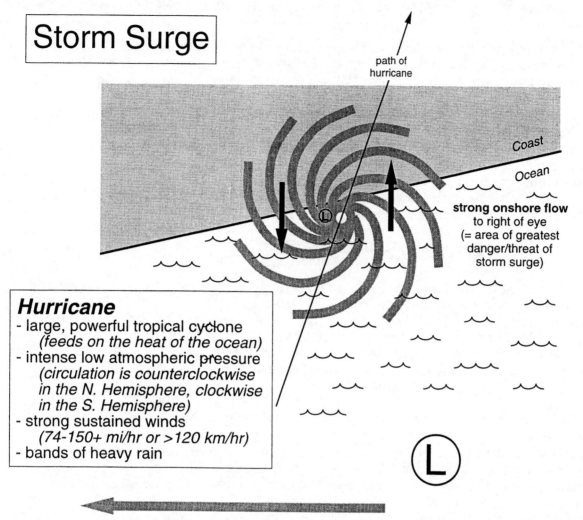

path of hurricane

Coast

Ocean

strong onshore flow
to right of eye
(= area of greatest
danger/threat of
storm surge)

Hurricane
- large, powerful tropical cyclone
  (feeds on the heat of the ocean)
- intense low atmospheric pressure
  *(circulation is counterclockwise
  in the N. Hemisphere, clockwise
  in the S. Hemisphere)*
- strong sustained winds
  (74-150+ mi/hr or >120 km/hr)
- bands of heavy rain

L

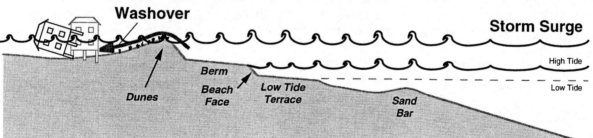

Washover

Storm Surge

High Tide

Low Tide

Dunes

Berm

Beach Face

Low Tide Terrace

Sand Bar

Storm Surge
- water pushed strongly landward by winds
  and the forward direction of storm track
- surface of water is domed-up (drawn-up)
  under intense low pressure of the storm
  (17-20 ft. or 5-6 m)
- danger further compounded if landfall
  coincides with high tide

Sea Level Rise & Shoreline Retreat: A Real Threat

Global sea level has not remained constant through time. The most recent episode of continental-scale glaciation reached its maximum extent of ice-sheet growth 18,000 years ago. As a consequence, global sea level was lowered by 120-130 m (~400 ft.); the shoreline was out near the edge of the present continental shelf break. Although the rate of sea level rise since the last ice age has dropped off over the last 2000 years, global sea level is still rising today. Last century, global sea level rose ~15 cm (0.5 ft.), more in some places and less in others. This may not sound like much, but considering that so many people around the world live near the edge of the ocean on **low-lying, flood-prone coastal plains** (like the East Coast and Gulf Coast of the U.S., Bangladesh, and the Netherlands), a very small rise of sea level has the potential to translate into a major inundation of the sea (see diagram below).

Coastal erosion and **shoreline retreat** are increasingly becoming a threat to public and private property because of rising sea level. Rising sea level also means a greater threat of **storm surge** associated with both tropical (hurricanes) and extratropical storms (e.g., nor'easters). During storm surge, water and waves can breach the barrier dunes causing sand to be deposited in **washover fans** on the landward side of the dunes. This is an important process in shoreline retreat as coastal environments retain their original profiles but migrate ("roll") very gradually landward.

A genuine concern is that part of the observed trend of rising global sea level may be attributed to **global warming** through the melting of ice sheets still covering Greenland and Antarctica, and/or to the thermal expansion of surface waters. Another contributing factor that may amplify the rate of relative sea level rise on a local or regional scale is **subsidence** (sinking) of the land, a problem that can be compounded by the extraction of pore fluids such as groundwater or hydrocarbons (oil and gas) near the coast. This has occurred around the Los Angeles area due to the pumping of oil from underground reservoirs. The city of New Orleans, which is at or below sea level, is vulnerable to storm surge associated with the direct hit by a major hurricane, or flooding along the Mississippi River. Levees built around the city to protect it from flooding have deprived the area of sediment that is needed to counteract the impact of subsidence of the land and rising sea level.

A reduction in the **supply of sediment** to the coastal zone can significantly contribute to the problem of shoreline retreat. Dams on our rivers provide vital hydroelectric power, flood control, and drinking water supplies, as well as recreation, but dams also deprive the coast of much needed sediment which becomes trapped in the reservoirs. Without mud, coastal wetlands, and particularly the salt marshes that border estuaries and bays may not be able to keep pace with rising sea level and drown, thereby contributing to the loss of these highly productive and vital coastal ecosystems. Without an ample supply of sand to the longshore transport system, there will not be enough material to maintain the natural give-and-take between the beaches, dunes, and offshore bars that occurs seasonally and protects the coast from flooding and erosion.

Rising sea level and shoreline retreat are but two of a number of challenges that many coastal communities, businesses, and individual property owners will face during the twenty-first century (see p. 184).

Sea Level Rise

2 examples with
sea level rise
of 0.3-1 m (1-3 ft.):
from the high tide shoreline
to the base of the dunes

Beach Profile

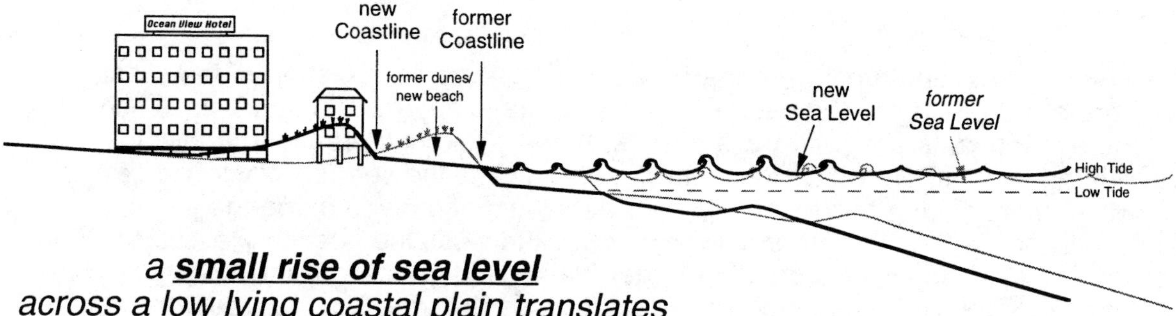

Ocean View Hotel
new Coastline
former Coastline
former dunes/ new beach
new Sea Level
former Sea Level
High Tide
Low Tide

*a **small rise of sea level***
across a low lying coastal plain translates
into a significant advance of the sea
*or **retreat of the shoreline***

coastal environments retain
their original profile
as they "roll" landward
with rising sea level

Barrier Island Profile

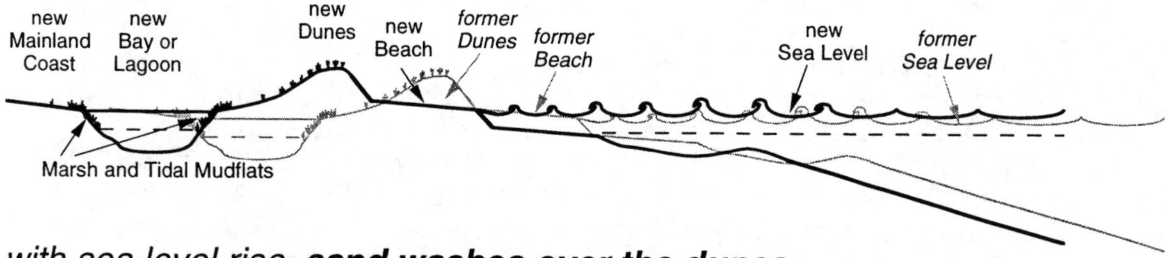

new Mainland Coast
new Bay or Lagoon
new Dunes
new Beach
former Dunes
former Beach
new Sea Level
former Sea Level
Marsh and Tidal Mudflats

*with sea level rise, **sand washes over the dunes***
and buries the marshes and mudflats, and
eventually fills in the bays/lagoons
*as the **shoreline retreats** and the*
beach and dunes migrate landward

Options in the Face of Rising Sea Level

Rising sea level, subsidence of the land, storm surge, and coastal erosion are growing threats to many coastal communities around the world. There are a number of different options available to protect coastal homes, businesses, and beaches. There are three general categories: 1) **hard stabilization**, or the "armoring" of the coast with rocks and manmade structures, is designed to protect the coast from erosion, trap sand, or redirect wave energy, 2) **soft stabilization**, also referred to as "beach nourishment", is intended to replace sand that has been lost from the longshore transport system, and 3) **relocation** of a structure or structures is an option that will preserve the natural character of the beach.

Seawalls, groins, jetties, and breakwaters are examples of hard stabilization of the coast, which have the effect of altering the natural dynamic of wave energy and sand movement along the shore. **Seawalls** are solid walls of rock, timbers, or concrete built parallel to the shore with the intent to protect buildings from washing into the sea (see facing page). Instead of wave energy being absorbed by the free movement of sand along the shore, the rigid seawall causes the wave energy to reflect back towards the ocean. Reflected waves collide with incoming waves creating larger waves and greater turbulence. Eventually, the walls will be undermined and fall over (see below). Seawalls destroy beaches over the long-term. Other examples of hard stabilization are summarized on the facing page.

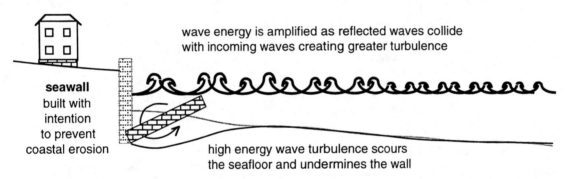

wave energy is amplified as reflected waves collide with incoming waves creating greater turbulence

seawall built with intention to prevent coastal erosion

high energy wave turbulence scours the seafloor and undermines the wall

Beach nourishment is the addition of sand to an eroding beach. Typically, this sand is dredged or pumped from an offshore location and put back onto the beach and back into the active longshore transport system. However, this process is temporary and costly. Several millions of dollars could be invested by a community, only to be lost in a single hurricane or nor'easter. Of the U.S. beaches that have been replenished, 26% last less than 1 year, 62% last 1-5 years, and ~12% last more than 5 years. Miami Beach is one of the best examples of a highly successful beach nourishment project.

Hard Stabilization of the Coast

longshore current

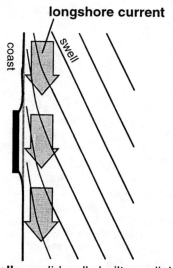

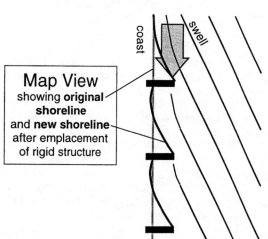

Map View
showing **original shoreline** and **new shoreline** after emplacement of rigid structure

seawalls - solid walls built parallel to shore

purpose: protect manmade structures and buildings from washing into the sea with rising sea level

result: incoming wave energy reflects off wall; reflected waves crash with incoming waves causing bigger waves and greater turbulence; eventually, walls will be undermined and fall over; seawalls destroy beaches over the long-term

groins - short structures built perpendicular to the shore, part-way across the surf zone

purpose: trap sand from longshore transport system and widen an eroding beach

result: sand will accumulate on the up-drift side and cause erosion on the down-drift side; may lead to construction of more and more groins down coast ("groin field")

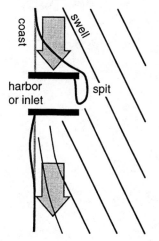

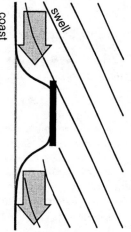

jetties - long structures built perpendicular to the shore.

purpose: protect harbor or channel inlet from incoming swell.

result: sand accumulates on the up-drift side and a sand spit will develop at end of jetty; must be dredged to prevent inlet from being filled with sand.

breakwaters - rigid structures built a short distance from, and parallel to, the shore.

purpose: provide quiet water for boat anchorage.

result: disrupts incoming swell which provides the energy to move sand via longshore drift; sand accumulates in quieter water behind the breakwater; must be dredged.

unconsolidated sand absorbs much of the ocean's wave energy in the surf zone
the free, unobstructed movement of sand along the coast protects the mainland from storms

Issues of Global Change: Food for Thought

An important goal of an introductory science course is to explore the "big picture" through "first principles"; in essence, to investigate the basic building blocks of understanding in a particular science. In this book, we have tried to provide a basic overview of the major features of the world ocean, as well as the atmospheric, biological, chemical, geological, and physical processes that make our Earth a habitable place to live. Hopefully, the exercises have also stimulated your interest in science and the workings of our home planet.

Science is important to all of us. Real environmental problems and difficult choices are confronting us. Issues of "**global change**" appear with greater regularity in the news; issues such as global warming, coral bleaching, El Nino intensity and frequency, rising sea level, storm surge, overfishing, pollution, and red tides. The burning of fossil fuels (coal, oil, and natural gas) over the past 150 years has unequivocally altered the composition of our atmosphere and scientific data now clearly demonstrate that global average temperatures are rising. In the 39 years between 1960 and 1999, the world's population doubled from 3 billion to 6 billion people. Nearly 20% of the world population lives in coastal cities, including 9 of the 10 largest cities, and >50% of the population lives within 100 km (62 miles) of the coast. Humanity will continue to exploit the sea, but its riches are not endless. The health of the world ocean should be of vital concern for all coastal nations. A major challenge for science is reaching a thorough understanding of our complex planet so that we can accurately forecast what the likely outcomes of our various activities will be over the long-term. For example, we do not yet know exactly how global warming will ultimately affect the coupled ocean-climate system. However, knowledge of oceanographic first principles will help guide us to sound, rational decisions concerning issues of global change. Here's some food for thought:

1. Overfishing
causes:
- vast improvements in fishing technology
  "too much fishing power chasing too few fish"
- indiscriminate fishing techniques
  "by-catch", biomass fishing
  taking of juveniles before reaching sexual maturity
- diminishing returns
  70% of the world's fish stocks are already being exploited at or beyond sustainable limits

potential impact:
- collapse of fisheries (too few breeding individuals to replenish the species)
  lose-lose situation for industry, consumers, and food webs

2. Degradation of coastal waters and loss of habitat
causes:
- pollution (raw sewage, toxic chemicals, heavy metals)
  poisoning of fish & shellfish by biomagnification
- agricultural runoff (fertilizers, pesticides) and nutrification
  algal blooms including "red tides", oxygen depletion, fish kills
- sediment runoff due to clear-cutting
  chokes the life out of coastal areas
- over-development of coast
  wetlands filled-in or modified

potential impact:
- serious damage to or loss of highly productive coastal waters ("fertile oases") which support coastal & oceanic food webs

3. Global warming
causes:
- burning of fossil fuels (oil, natural gas, and coal)
  source of CO and CO_2
- destruction/burning of forests
  instead of being a natural sink for CO_2, burning is source of CO and CO_2

potential impact:
- altering climate patterns (e.g. rainfall)
- greater extremes of weather
  floods, droughts, increased frequency and/or severity of El Nino events, hurricanes?
- rise of global sea level caused by thermal expansion of seawater and melting ice
  coastal flooding, greater risks of storm surge

Index